P9-DDY-663

SECRET EMPIRES

ALSO BY PETER SCHWEIZER

Clinton Cash: The Untold Story of How and Why Foreign Governments and Businesses Helped Make Bill and Hillary Rich

Extortion: How Politicians Extract Your Money, Buy Votes, and Line Their Own Pockets

Throw Them All Out: How Politicians and Their Friends Get Rich off Insider Stock Tips, Land Deals, and Cronyism That Would Send the Rest of Us to Prison

Architects of Ruin: How Big Government Liberals Wrecked the Global Economy—and How They Will Do It Again If No One Stops Them

Landmark Speeches of the American Conservative Movement (with Wynton Hall)

The Reagan Presidency: Assessing the Man and His Legacy (with Paul Kengor)

Do as I Say (Not as I Do): Profiles in Liberal Hypocrisy

The Bushes: Portrait of a Dynasty (with Rochelle Schweizer)

Reagan's War: The Epic Story of His Forty-Year Struggle and Final Triumph over Communism

Disney: The Mouse Betrayed (with Rochelle Schweizer)

Victory: The Reagan Administration's Secret Strategy that Hastened the Collapse of the Soviet Union

SECRET EMPIRES

How the American Political Class Hides
Corruption and Enriches Family and Friends

PETER SCHWEIZER

HARPER
An Imprint of HarperCollins*Publishers*

HarperCollins books may be purchased for educational, business, or sales promotional use. For information, please email the Special Markets Department at SPsales@harpercollins.com.

FIRST EDITION

Designed by William Ruoto

Library of Congress Cataloging-in-Publication Data has been applied for.

ISBN 978-0-06-256936-3

18 19 20 21 22 LSC 10 9 8 7 6 5 4 3 2 1

For

Jack & Hannah

CONTENTS

SECRET EMPIRES

1

CORRUPTION BY PROXY

> Public office is a privilege, not a right, and people who accept the privilege of holding office in the Government must of necessity accept that their entire conduct should be open to inspection by the people they are serving.[1]
>
> —HARRY TRUMAN

- A new corruption that is extremely difficult to detect is taking over our political system.
- It is most lucrative; instead of thousands of dollars, the sums amount to millions or even billions.

The 2016 election left Americans deeply divided. Some found themselves enthused and encouraged by the outcome. Others were disappointed and depressed. But the vast majority of the American people were unified about one thing: the belief that corruption is rampant in Washington. Polls revealed that three out of four Americans believe that there is "widespread government corruption" in Washington, and perhaps the holdouts are not paying attention. The concern over widespread corruption is growing. The number was two in three back in 2009.[2]

Politicians who practice corruption have created a crisis of confidence in our government. Only 19 percent of the American people in a recent survey trust the federal government to do the right thing.[3]

Americans have known for quite some time that some politicians make themselves rich through public service. Who could forget Louisiana congressman William Jefferson stashing $90,000 in bribes in his freezer?[4] Or Congressman Randy "Duke" Cunningham who created a "bribery menu" that netted him $2.4 million?[5] Then there was the 2006 Jack Abramoff bribery case that led to a flurry of convictions in Washington. The lobbyist had given gifts to politicians in exchange for favors. Abramoff was sentenced to four years in jail.[6] Congressman Bob Ney pleaded guilty to taking bribes from Abramoff and was sentenced to thirty months in jail.[7] Congressional aides for no less than five congressmen pleaded guilty to corruption, conspiracy, obstruction of justice, and bribery charges, as did executives in the Bush administration from both the Department of the Interior and the Justice Department.

Other forms of corruption have been exposed over the past decade that, while not leading to jail time, have led to great public scrutiny and embarrassment. In my book *Throw Them All Out*, I exposed insider trading on the stock market by members of Congress. When the book was published and featured on *60 Minutes*, it set off a national firestorm that led to the passage of the Stop Trading on Congressional Knowledge (STOCK) Act.[8] In my book *Extortion*, and in another *60 Minutes* story, I revealed how politicians enriched themselves using mafia-like tactics.[9] In *Clinton Cash*, I laid out how the Clintons monetized access and official favors.[10] Polls conducted in 2015 and 2016 showed that a sizable portion of the American public believed that the Clintons were corrupt.[11] And the Federal Bureau of Investigation launched an investigation based on the book's findings.[12]

Following the money in those projects was simpler, using financial disclosures required by law.

The corruption by proxy that is the subject of this book is far more troubling and difficult to detect. Because the transactions involved do not fall under disclosure laws, they are invisible or, at least, hidden from public scrutiny. Financial deals channeled through politicians' family members do not require disclosure. Identifying deals and the parties involved takes intense research. These transactions nevertheless make good money for a politician and his family and friends. Politicians are constitutionally obligated to make decisions that are best for the people they serve. These deals direct the politician's loyalty elsewhere.

We are used to the typical "revolving door" corruption where government officials will carry out policies and then, after leaving office, take a job from those who benefited. But with most of the deals covered in this book, these politicians do not wait until they leave office. The accumulation begins while they are still serving. Rather than personal accumulation, the wealth flows to someone who is important in their life—a child, another family member, or a close friend. Those relatives are not required to disclose publicly how much money they are making or from whom. It is a pernicious form of what I call corruption by proxy.

Politicians are like the rest of us in that they avoid overtly criminal or publicly embarrassing behavior. Today, only the most arrogant, desperate, or stupid politician would take a direct payment from someone in exchange for a favor. Even without a clear exchange, savvy politicians are likely to avoid transactions that lead to a dramatic increase in their own net worth, lest they draw negative attention from an already skeptical public.

So a new corruption has taken hold. It involves larger sums of money than ever before. Instead of stuffing tens of thousands of dollars into a freezer, they stuff multibillion-dollar equity deals done in the dark corners of the world into family members' bank accounts.

As some political leaders are making influential decisions that affect everything from national security to our standard of living, large sums of money are funneled to their families and closest friends. Only the most naive would believe that these deals are a coincidence. This money often flows from foreign governments or oligarchs. The timing of these transactions and the amounts involved raise serious questions about the leaders' integrity and decision making. The high-level policy choices make the news, but the transactions quietly taking place below the surface do not.

How do I define corruption? By corruption, I mean when officials "abuse their positions for private gain . . . Corruption is the dishonest or fraudulent use of power for personal gain."[13]

It is often illegal—but not always. Some forms of what most of us would regard as corruption are still technically "legal."[14] These cases, where abuse of power is most certainly occurring, and where the public appears to have no legal recourse short of voting the abuser out of office, are especially demoralizing to Americans. In the old days of the Tammany Hall machine, this was known as "honest graft." George Washington Plunkitt, who helped run the corrupt Tammany machine, explained that often what he called honest (legal) forms of corruption were more profitable than the illegal forms. "There is so much [honest graft] in this big town that they would be fools to go in for dishonest graft," he famously said.[15]

In the world of finance, everyone understands the subterfuge of "offshore" assets. Corporations and high-net-worth individuals stuff their assets and money into obscure limited liability companies or bank accounts in places like Panama, Belize, and Cyprus. By doing so, it becomes very difficult, if not impossible, to map out their transactions. Some do it to avoid taxes. Others try to obscure how and with whom they do business.

When it comes to corruption in Washington, the same sort of thing is occurring. The American political class is "offshoring" its corruption.

Sometimes, politicians will literally offshore corruption by moving the location of the deal, the entities, and the players involved, abroad. Other times, they offshore corruption metaphorically by shifting the cronyism out of their own hands and into those of their children or a close friend to leverage from afar and avoid detection. Either way, the results are the same: like the financier who puts his assets in a bank in Belize, it becomes very hard to track the flow of money and exchange of services. Politicians and their families are increasingly able to avoid public scrutiny and accountability for the selling of influence. And the sums involved are enormous.

So exactly how are politicians "offshoring" their corruption?

First, the foreign entities that are providing the funds and favors for politicians oftentimes have fewer disclosure rules than those in the United States. Again, to use the analogy of offshore banking, the financier who is trying to hide his assets would rather put his money in the Bank of Belize than in the Bank of America. The Bank of Belize has far fewer disclosure rules than the Bank of America. Corrupt politicians like this. A Chinese firm providing deals for them or their family is not required to disclose much of anything when compared to a U.S. publicly traded company. So in a strange way, corrupt politicians would rather seek financial opportunities from shady foreign entities than American corporations. Bigger paydays with less scrutiny result.

Second, by offshoring their corruption via deals for their family and close friends, politicians are not required to disclose such financial gains. In American politics, great scandals have fortunately led to ethics crackdowns. The Bobby Baker scandal, which rocked the U.S. Senate, exposed influence peddling, tax evasion, fraud, and theft by an influential Senate aide. This led to the creation of a Committee on Standards and Conduct in 1964. The Senate created its first ethics code four years later following another financial scandal involving Senator Thomas Dodd. In 1991, a scandal involving then senator David Durenberger forced the Senate to ban members from receiving speaking fees. In the House, a scandal involving Adam Clayton Powell Jr. led the House to establish a Committee on Standards of Official Conduct in 1967 and to pass ethics codes in 1968 and 1970.[16]

As a positive result of these crackdowns, politicians have been required to disclose their income, assets, and financial transactions, as well as those of their spouses. In short, if a politician or a politician's spouse directly gets a big payday, it would be illegal to hide. Failing to report it would be a violation of federal law. Yet failing to report that their adult child or friend got a big payday as a result of the politician's policy actions is not against any law. It is "offshore." None of those jackpots needs to be disclosed to the public.

Besides being off the disclosure grid, offshore transactions provide plausible deniability when facing scrutiny. If money lands in a politician's bank account, there is evidence of a direct benefit. Even if it shows up in the bank account of their spouse, questions would be raised. But what if the money, the job, or the business opportunity falls to a politician's adult child or best friend? Even if that transaction were to become public, there would be an added layer of legal protection.

In short, this is corruption by proxy. It is essentially a form of "political arbitrage," where friends and family members of powerful political figures have positioned themselves to serve as conduits or middle men between those seeking influence and those who possess political power. This creates previously unimaginable pathways to wealth for the political class. Better still for them, because of limited disclosure rules, much of it can go on undetected.

Foreign governments and oligarchs like this form of corruption because it gives them private and unfettered gateways to the corridors of Washington power. There are a myriad of things they may want: approval to take over an American company; the transfer of sensitive technology; trade deals; or favorable policies concerning military or national security matters. Foreign entities cannot legally make campaign contributions, so using this approach creates an alternative way to curry favor and influence America's political leaders. Simply camouflaging these transactions as business agreements provides another shield of plausible deniability.

Such corruption is especially bad because it makes American politicians vulnerable to overseas financial pressure. The political leaders identified in the chapters to follow have often seen themselves and their families become wealthy with a single deal provided by a foreign government. So we have American politicians whose wealth is tied up with being in the good graces of foreign governments. What could possibly go wrong?

Using a politician's adult child or close friend as a proxy may provide only an indirect benefit to a politician, but that does not make it less troubling or threatening to the public interest. Indeed, the law says that if an elected official performs an act for someone's benefit and that someone provides a benefit to their family or close friend—it is still illegal. Just as a foreign bank account set up under a foreign limited liability company does not shield the financier from scrutiny for trying to avoid taxes, proxy corruption should not limit scrutiny of the politician or his shell game.

Make no mistake: enriching a friend or family member of a politician and getting something in return still constitutes a bribe. The Organisation for Economic Co-operation and Development (OECD) is an international body of the world's largest economies. As the OECD's Anti-Bribery Convention notes, "individuals and companies can also be prosecuted when third parties are involved in the bribes transaction, such as when someone other than the official who was bribed received the illegal benefit, including a family member, business partner, or a favourite charity of the official."[17]

Direct bribery is only the most obvious form of corruption. Politicians (at least those who are halfway smart) avoid that sort of behavior. It has long been recognized that direct bribes are the least of our concerns when it comes to corruption. Even a century ago, a scholar noted that there are far more troubling and sophisticated forms of bribery. As Professor Robert Brooks wrote back in 1909, "It is this condition of affairs which makes the subtler aspects of corruption so much more dangerous and so much less easy to cope with than common bribery."[18]

Brooks also noted that the smart politician of his day would avoid obvious bribery. "Corruption in its more insidious forms is not the vice of low intellects," he noted.[19] And so it still is today. Corruption by proxy takes some sophistication.

AMERICAN PRINCELINGS

Throughout human history, almost regardless of culture, the family has proved to be an effective gateway to curry favor with the powerful. In Europe, Africa, and Asia, high-ranking political figures and oligarchs have used marriage and family as a means of gaining political standing and influence. European kings would marry off their sons and daughters to rivals or competitors, thereby striking a bargain or alliance. It was an efficient—if less than romantic—means of accumulating power and wealth. In Europe, the royal houses often intermarried out of political expediency. Famously, Mary, Queen of Scots, was crowned after her father James V died while she was but an infant. After a betrothal to Henry VIII's infant son, Edward, failed to ameliorate tension with England, she was sent to France, her mother's homeland, as a five-year-old. There she was betrothed to marry Francis, the heir to Henry II, king of France. After the ineffectiveness of the first betrothal, it was part of what was hoped to be a Franco-Scottish alliance against England.[20]

In the United States, these arrangements have also been historically true, even if the stories lacked the titles of kings, queens, and princes. American politicians from the earliest days would marry their children off to the wealthy, leading to some powerful alliances. As Stephen Hess has pointed out in his classic study *America's Political Dynasties*, "the correspondence of Presidents John and John Quincy Adams is filled with tales of money miseries; then the latter's son married the daughter of Peter Chardon Brooks, Boston's first millionaire."[21] The powerful Livingston family of New York in one generation had five girls marry U.S. congressmen, creating a powerful commercial-political alliance. The great industrialists of the nineteenth century often married their daughters off to politicians so they could literally be "wedded" to power in Washington.[22]

Of course, today such arranged marriages would be considered outrageous. The notion of an American politician offering his daughter to a Chinese oligarch in an arranged marriage for convenience or alliance offends our modern sensibilities. But such "marriages" still take place. Instead of taking the form of matrimony or vows of fidelity, the union instead comes in the form of private equity or business alliances. The tie that binds is not marriage but money. The cross-border marriage has been replaced by the cross-border joint venture. Foreign oligarchs are eager to "marry" into notable American political families via lucrative financial deals, thereby leveraging access to the highest reaches of political power.

The alliance of American political family members with foreign powers is a relatively new phenomenon in the United States. Certainly, there have been occasional examples that were considered scandalous behavior in the past. Billy Carter, brother of President Jimmy Carter, famously took money from the Libyans.[23] President Bill Clinton's brother Roger took money from the South Koreans.[24] President George W. Bush's brother Neil got a nice deal from a Chinese computer company backed by the son of the Chinese president.[25]

These alliances seem essentially ad hoc and haphazard, and the amounts of money small when compared to the alliances revealed in this book. Instead of hundreds of thousands of dollars, we are now seeing hundreds of millions of dollars involved in calculated unions that exploit American political family ties.

It is becoming more commonplace, more lucrative, and more systematic for the children of powerful American politicians to go into highly profitable business because people want to curry favor with their important parents.

This is, after all, the way corruption often occurs in other countries. In almost every country, family ties are exploited for financial gain. In many parts of the world, the children of powerful political figures go into business and profit handsomely, not necessarily because they are good at business, but because people want to curry favor with their influential parents. In China, for example, hundreds of children of high-ranking Communist Party officials have gone into business over the decades. They thrive precisely because someone wants something from their powerful parents. They even have a term for them: the Princelings.[26]

As we will see, many of the deals discussed here involve Chinese state-owned companies doing deals with family members of America's most senior politicians. China employs a strategy with other countries to make friends with foreign officials and politicians to advance their interests.[27] In Canada, the head of the Canadian Security Intelligence Service noted a few years ago how Chinese officials were "cozying up to politicians" to score favorable policies. A Canadian member of Parliament went on national television and described how when his colleagues traveled to China they were offered lucrative business deals by Chinese officials.[28]

China, of course, is not alone in doing this. As we will see, other countries employ this strategy as well. And they largely do it through the family members of politicians because that is the way such deals are done in their own countries.

The problem, of course, is not commercial deals with China per se. I believe that business and trade relationships between American and Chinese firms are healthy. The problem is commercial and financial deals between Chinese government entities and the families of American politicians. It is hard not to see these sorts of deals as sophisticated payoffs or bribes.

Part of the reason that China figures so prominently is because the world's most populous country is one of the fastest-growing economies on the planet. There is lots of money to be made and passed around. China also boasts a political culture that is comfortable with juicing "Princelings" into deals, which is one of the reasons Sarah Chayes, in her book on global corruption, considers it "one of the world's most corrupt countries."[29] But China is not alone: countries like Ukraine and Russia also figure prominently in this story because they also have political and economic cultures that are deeply corrupt. And America's top political families appear all too eager to sign fruitful deals with them.

The American media has run endless stories on the Chinese Princelings. But as we will see, American Princelings are the sons, daughters, siblings, and closest friends of America's most powerful political figures—presidents, vice presidents, cabinet officials, senators, and congressmen—who have accrued financial benefits because of the political power that their parents, family, or friends wield. And it is not just federal officials who have been attracted to the Princeling method. Even mayors have embraced it, striking alliances with corrupt officials halfway around the world.

Foreign governments and oligarchs are seeking personal financial relationships with America's political elite. And many of those elite appear all too happy to oblige, their increased wealth perpetuating their family empire's elite status. America's political elite thus focus their loyalty and obligation on their kids' financial interests, or the family financial "empire," over loyalty owed to their country. While we admire people, including politicians, who prioritize love and support of their families, especially their children, we must also clearly hold them as public servants to a standard that does not compromise our nation by abusing their vested power on behalf of family and friends.

GLOBALIZATION OF CORRUPTION

The explosive growth in global business markets has changed the face of corporate America. As American corporations have become global economic players, they have lost much of their American identity and loyalty. Back in 1996, Ralph Nader wrote to the chief executives of one hundred of America's largest corporations. After pointing out the fact that these corporations received substantial tax benefits and subsidies from the United States, he urged them to show support for "the country that bred them, built them, subsidized them, and defended them" by opening their annual stockholders' meetings with the Pledge of Allegiance. Only one corporation responded favorably (Federated Department Stores). The others dismissed the idea on the grounds that, in the words of the Ford Motor Company, they are "multinational," not American.[30]

It has been argued that globalization leads to greater world peace by the blurring of national borders. Commercial relationships between countries and people cause them to share more common interests and become more cooperative, so the reasoning goes. But as commercial relations bring countries closer, they will certainly bring political elites together. What happens when American political elites and their families create strong, that is lucrative, commercial ties with foreign governments? Compromised national security does not equate with world peace.

As our business markets have become globalized, so too has corruption. Globalization is blurring how our politicians view their first allegiance to country. Those exposed in this book would all vehemently deny that their families' commercial deals with foreign governments and oligarchs influence their decision making or make them sympathetic to the political needs of their commercial partners. Not only does this seem preposterous on its face, but by doing so they are denying a central tenet of the commercial globalism that they endorse. Free trade and exchanges between nations generally create goodwill between countries, but profitable commercial deals between foreign governments and America's political elite undermine the latter's vision for leadership by warping their priorities.

American corporations have broadened their allegiances because their customers and operations are increasingly global. In other words, their financial interests extend far beyond America's borders. The same is happening to some American politicians. Those whose families strike profitable commercial deals with foreign governments or corporations can naturally be expected to show greater sympathy to those foreign entities. But how will they handle issues when the interests of their foreign commercial partners collide with the interests of the United States? Foreign governments looking for favorable treatment from the powerful in Washington easily attach financial strings in the guise of "globalization."

Foreign governments and oligarchs have become more aggressive in Washington over the past half-century. This coincides with foreign governments spending more and more to court our politicians. For example, the late Harvard professor Samuel P. Huntington allows that in "the mid-1980s Mexico was spending less than $70,000 a year on lobbying Washington, and President Miguel de la Madrid (a graduate of the Harvard Kennedy School of Government) lamented the difficulty he had getting his diplomats not just to deal formally with the State Department but to develop close relations with the congressmen who had the real power to affect Mexico's interests."[31] Today, the government of Mexico and Mexican companies, along with many other countries, hire armies of lobbyists and government-relations operatives to get what they want from Washington.[32] Foreign governments who cultivate commercial ties directly with our politicians and their families have realized that this is a more efficient investment.

At a minimum, these sorts of deals make foreign oligarchs and governments unofficial constituencies that politicians are incentivized to please, thanks to these economic ties.

The bottom line is that globalization has transformed the world economy, but it has also, in the words of Laurence Cockcroft in his book *Global Corruption*, "both exposed and accelerated corruption."[33]

SMASH AND GRAB

It has been said that to the extent that government can do something for you, it can also take something from you.[34] And sometimes nearly destroying an industry or business can create opportunities for a powerful politician's friends, who create wealth by proxy, and can then channel some of it back to the politician's interests or war chest.

We traditionally think of cronyism as politicians doing favors for well-heeled allies and friends. Maybe it takes the form of corporate welfare—that is, providing government grants or favors to connected businesses. Cronyism could take the form of government bailouts for a failing business or perhaps government-backed loans. In short, traditional cronyism uses the power of government to provide commercial opportunities or favors for connected friends and family members.

But there are other favors that the government can perform. In recent years a new method of cronyism has emerged that we will call "smash and grab."

Ask any street cop and he or she will describe the crime this way: someone sees a wallet or a purse sitting on the front seat of a locked car. The culprit wants it. So he smashes the window, grabs the wallet or purse, and takes off.

Smash and grab in government works in a similar way, only while one guy smashes, another grabs. Say there is a particular company or industry with large assets. The government, by their words or policies, "smashes" the industry on the grounds that it is bad, destructive, or dangerous. This is often done because an industry or company is deemed harmful to the environment or damaging to public health, or it exploits vulnerable people. Once "smashed," the valuation of that industry or company drops dramatically. But then something else happens. Investors or financiers closely tied to that politician suddenly buy the company or buy into the industry for pennies on the dollar. The company or industry is then resurrected to its previous luster and its valuations rise dramatically under new owners who have close ties to the politicians.

Smash and grab is an effective form of cronyism because, as with Princelings, it is by proxy and thus hard to track. Unlike traditional forms of corporate welfare, the action of smashing a company or an industry is often wrapped in a benevolent light. The very act of smashing an unpopular industry is presented to the public as a virtue, not as a vice. This, of course, can obscure the larger crony pattern.

As we will see, in recent years the federal government has smashed a variety of industries—coal, for-profit colleges, oil companies, payday lenders—while politically connected financiers who are close to those doing the smashing have then swooped in and bought up those same companies at a deep discount. The politician gets to claim a virtuous act done in the public's interest, while his or her closest friends and money supporters clean up financially. Smash and grab is profoundly destructive because it not only manipulates the economy, but also generates wealth for the close friends of politicians by first destroying someone else's wealth.

BEYOND DISCLOSURE

With the public keenly resentful of and focused on traditional corruption, these new proxy methods have become increasingly popular with politicians. When corruption is detected and exposed, it has consequences. Recent political history demonstrates that people are very concerned about political corruption and leaders lose power when it is exposed. For example, a series of scandals in 2005 and 2006 resulted in the resignation of four Republican members of Congress. The Republicans lost their majority status in the 2006 election.[35]

In 2016, persistent questions about Bill and Hillary's financial transactions in light of her duties as secretary of state, as well as the scandal involving her secret e-mail server, damaged her popularity with voters and helped cost her the election.

Much of the media focus on the question, "What are politicians required to disclose?" They track individual campaign contributions, donations from political action committees, and independent expenditures for campaigns. Transparency is essential in these areas and requires ongoing public vigilance to be effective. Yet there is so much more hiding offshore. While traditional forms of corruption continue to occur, many politicians use proxies to cloak their transactions.

Our legal system ends up unevenly applying justice and failing to prohibit the corrupt practices that most threaten sound government. Think about it: failing to disclose a $2,700 campaign contribution, or making a quid pro quo deal for that donation, will land a politician in legal trouble. But what about a multibillion-dollar private equity deal involving a family member and foreign government? While the $2,700 contribution must be disclosed by law, the billion-dollar deal does not need to be reported. And while trading favors for a campaign contribution can get you sent to jail, the legal system essentially says about a private equity deal, "Nothing to see here. Move along."

In earlier eras, corruption was mostly limited to individuals exploiting their positions of power and influence. It could be beaten back, if not completely expunged. Today's corruption is different—harder to see, financially complex, and therefore harder to combat. Most interestingly, perhaps, the political class possesses an ideology that in a true Marxian sense serves to justify its authority and privilege in its own eyes. Its members would have us see no conflict between their self-interest and the public good. Its members would have us believe that those private deals will not influence their judgment.

"Trust us," they say.

Boss Tweed, who ran the Tammany Hall operation, used the typical bandwagon defense: "The fact is New York politics were always dishonest—long before my time. There never was a time when you couldn't buy the Board Alderman. A politician coming forward takes things as they are. This population is too hopelessly split up into races and factions to be governed by universal suffrage, except by bribery of patronage or corruption."[36] You hear echoes of the same in Washington today. Many political elites see inside deals as the oil that keeps the machinery of government working, and keeps smart people like them in public office. It is dishonesty in the name of empire-building.

Others will argue that corruption really does not matter; that it is a victimless crime. So what if politicians and their families get rich? Who cares? It doesn't affect me.

But corruption plays a profound role in the kind of leadership we get, because corruption is upstream from governing.

A spill of toxins upriver from where you are drinking will foul the water you consume—and even make it poisonous. Those who believe that corruption does not really affect them fail to see that corruption is not a by-product or an afterthought in bad policies. It pollutes policies even as they are fashioned.

As we will see, vast amounts of toxins are being dumped upstream.

Machiavelli in his classic work *The Prince* offered blunt advice to those interested in political power. He declared that it was a good idea to be mean and ungenerous. He encouraged political figures to be harsh and to show no mercy. These were quality characteristics that could help someone stay in power. But there was one vice that he admonished his readers to avoid: corruption. "Being rapacious and arrogating subjects' goods and women is what, above all else . . . renders him hateful." Machiavelli declared that it was better to be feared than loved as a leader, but you did not want to be hated. Hatred led to conspiracy. And conspiracy, he warned, could bring down governments.[37]

2

AMERICAN PRINCELINGS: TWO SONS AND A ROOMMATE

- The son of Vice President Joe Biden and the stepson of Secretary of State John Kerry profited immensely from secretive deals they struck with companies connected to the Chinese government.
- These deals occurred as Biden and Kerry were negotiating sensitive issues with the Chinese government.
- The deals involved billions of dollars.

Joe Biden and John Kerry have been pillars of the Washington establishment for more than thirty years. Biden is one of the most popular politicians in our nation's capital. His demeanor, sense of humor, and even his friendly gaffes have allowed him to form close relationships with both Democrats and Republicans. His public image is built around his "Lunch Bucket Joe" persona. As he reminds the American people on regular occasion, he has little wealth to show for his career, despite having reached the vice presidency.

One of his closest political allies in Washington is former senator and former Secretary of State John Kerry. "Lunch Bucket Joe" he ain't; Kerry is more patrician than earthy. But the two men became close while serving for several decades together in the U.S. Senate. The two "often talked on matters of foreign policy," says Jules Witcover in his Biden biography. When John Kerry was the Democrat nominee for president in 2004, Biden had been on the "short list" as his running mate. It was widely believed that if Kerry had won in November, he would have selected Biden as his secretary of state.[1] So their sons going into business together in June 2009 was not exactly a bolt out of the blue.[2]

But with *whom* their sons cut lucrative deals while the elder two were steering the ship of state is more of a surprise.

What Hunter Biden, the son of America's vice president, and Christopher Heinz, the stepson of the chairman of the Senate Committee on Foreign Relations (later to be secretary of state) were creating was an international private equity firm. It was anchored by the Heinz family alternative investment fund, Rosemont Capital.[3] The new firm would be populated by political loyalists and positioned to strike profitable deals overseas with foreign governments and officials with whom the U.S. government was negotiating.

Hunter Biden, Vice President Joe Biden's youngest son, had gone through a series of jobs since graduating from Yale Law School in 1996. First, he had taken a position as an executive with MBNA, a major credit card–issuing bank based in Delaware, the state his father represented in the U.S. Senate. The young Biden handled legal matters for MBNA, one of his father's largest campaign contributors. He then moved on for a brief stint in the Clinton administration at the Department of Commerce, serving under Bill Daley, brother of Chicago's mayor.[4] By 2001 he joined with William C. Oldaker, a longtime Washington, D.C. hand and campaign adviser to his father, to open a lobbying shop.[5] The law firm of Oldaker, Biden, and Belair, with offices on Connecticut Avenue (just blocks from the White House), represented clients including online gambling firms, universities, and Native

American tribes.[6] Hunter stayed in the firm until 2008, when having a son as a lobbyist apparently did not cast the right image for his father, who had been picked as Barack Obama's running mate.[7]

Hunter had tried other ventures. In 2006, he joined forces with his uncle, James Biden, to dive into the hedge fund business. The two purchased an investment firm called Paradigm Capital Management, which reportedly had $1.5 billion under management.[8] It was founded by Dr. James Park, who happened to be the son-in-law of Rev. Sun Myung Moon, head of the Unification Church.[9]

Hunter Biden took the helm as chairman, but things did not go well with Paradigm almost from the beginning. The Bidens quickly discovered the firm didn't have $1.5 billion under management; it had far less than half that.[10] And the news got worse. A tenant named Ponta Negra Group that was subleasing an office from Paradigm was charged with orchestrating a multimillion-dollar investment fraud.[11] Paradigm also became associated with Stanford Capital Management, which was exposed as a giant Ponzi scheme in one of the largest cases of financial fraud in American history.[12] Paradigm managed a fund for Stanford Capital but was never accused of any wrongdoing. The Bidens did end up in major litigation involving charges of fraud and misrepresentation regarding their purchase of Paradigm. The Bidens insisted that they did nothing wrong and settled the suit in December 2008.[13]

Much of Hunter Biden's professional career followed in the shadow of his father's trajectory. And he, along with the rest of the Biden family, "formed the nucleus for [Joe Biden's] political operations."[14] Politics was a closed loop, a family affair. Hunter's job during his father's many campaigns over the years included standing below the podium while Biden was speaking and pointing to his watch to remind him when to stop talking.[15] And the Bidens mixed politics, family, and business on basic levels. When Joe Biden ran for office, he threw campaign dough in the direction of his family members. By 2008, Biden's campaigns had paid more than $2 million to family members and their businesses.[16]

By their own account, Joe and Hunter are extraordinarily close. "The single best thing [I learned from my father]," Hunter once said, "is family comes first. Over everything."

"We have an expression in our family," says Vice President Biden. "If you have to ask for help, it's too late. We're there for each other."[17]

By 2009, Hunter's father was the vice president of the United States.

With Joe Biden arguably the second most powerful man in the world, Hunter Biden became a social fixture in Washington, photographed regularly with the powerful and the beautiful. He and his then wife, Kathleen, mourned with his father and mother at Arlington National Cemetery when Senator Edward Kennedy was laid to rest in August 2009.[18] In November 2009, Hunter secured a prized invitation to the state dinner for the visit of India's prime minister.[19] He took in the Georgetown University versus Duke University basketball game at the Verizon Center in Washington. He had courtside seats—next to his father and Barack Obama.[20] He also participated in celebrity events like one organized by musical star Usher's foundation.[21]

By the summer of 2009, the thirty-nine-year-old Hunter joined forces with the son of another powerful figure in American politics, Chris Heinz. Senator John Heinz of Pennsylvania had tragically died in a 1991 airplane crash when Chris was eighteen. Chris, his brothers, and his mother inherited a large chunk of the family's vast ketchup fortune, including a network of investment funds and a Pennsylvania estate, among other properties.[22] In May 1995, his mother, Teresa, married Senator John Kerry of Massachusetts.[23] That same year, Chris graduated from Yale, and then went on to get his MBA from Harvard Business School.[24]

Joining them in the Rosemont venture was Devon Archer, a longtime Heinz and Kerry friend.[25] He had been roommates with Chris at Yale and then proved himself to be more than Chris's wingman by emerging as a star fund-raiser for Senator John Kerry's 2004 presidential campaign. By aggressively using his contacts, he helped raise millions of dollars for the race and, at the ripe age of thirty, landed a spot as the cochairman of the Kerry campaign's finance committee. It was an impressive achievement.[26]

Archer was tenacious. "If you're a real doer," he told the *Observer* at the time, "you get respected, and that respect is reflected in how the campaign treats you, because there are a ton of people out there who are a lot of hot air."[27]

Archer proved he was anything but hot air. After the campaign, he went on to become a trustee of the Heinz Family Office, a sprawling network of investment funds worth close to $1 billion, of which Rosemont Capital is a part. He was also selected to join the board of the family's Howard J. Heinz Trust.[28]

So the three friends Biden, Heinz, and Archer, established a series of related LLCs.

The trunk of the tree was Rosemont Capital, the alternative investment fund of the Heinz Family Office. Rosemont Farm is the name of the Heinz family's ninety-acre estate outside Fox Chapel, Pennsylvania.[29] The small fund grew quickly. According to an e-mail revealed as part of a Securities and Exchange Commission (SEC) investigation, Rosemont described themselves as "a $2.4 billion private equity firm co-owned by Hunter Biden and Chris Heinz," with Devon Archer as "Managing Partner."[30]

The partners attached several branches to the Rosemont Capital trunk, including Rosemont Seneca Partners, LLC, Rosemont Seneca Technology Partners, and Rosemont Realty.

Chris Heinz served as the managing director and cochairman of Rosemont Capital. (In 2014, he stepped down as cochairman, but remained a founding partner, vested in the firm.)[31] Devon Archer was the cofounder and senior managing partner at Rosemont Capital. Hunter Biden was the managing partner at Rosemont Seneca Partners.

Of the various deals in which these Rosemont entities were involved, one of the largest and most troubling concerns Rosemont Seneca Partners.

Rather than set up shop in New York City, the financial capital of the world, Rosemont Seneca leased space in Washington, D.C. They occupied an all-brick building on Wisconsin Avenue, the main thoroughfare of exclusive Georgetown. Their offices would be less than a mile from John and Teresa Kerry's twenty-three-room Georgetown mansion, and just two miles from both Joe Biden's office in the White House and his residence at the Naval Observatory.[32]

Over the next seven years, as both Joe Biden and John Kerry negotiated sensitive and high-stakes deals with foreign governments, Rosemont entities secured a series of exclusive deals often with those same foreign governments. Some of the deals they secured may remain hidden. These Rosemont entities are, after all, within a *private* equity firm and as such are not required to report or disclose their financial dealings publicly.

Some of their transactions are nevertheless traceable by investigating world capital markets. A troubling pattern emerges from this research, showing how profitable deals were struck with foreign governments on the heels of crucial diplomatic missions carried out by their powerful fathers. Often those foreign entities gained favorable policy actions from the United States government just as the sons were securing favorable financial deals from those same entities.

Nowhere is that more true than in their commercial dealings with Chinese government–backed enterprises.

For U.S. financial firms, the gold rush for the past two decades has been business in China. The opportunities appeared endless in the most populous country, with one of the highest growth rates, in the world.[33] So the Rosemont team set their sights on the Middle Kingdom.

Effective diplomacy, according to Joe Biden, is about forming personal relationships with foreign leaders. "It all gets down to the conduct of foreign policy of being personal."[34]

Doing business in China often entails having the right contacts and relationships; simply having the right family name can be an enormous benefit to getting deals. Unlike in the West, in China, surnames appear first.[35] Political connections are especially important: despite outward signs of a freewheeling capitalist economy, Chinese government officials still wield great power that can make or break a deal.

To jump-start deals, some financial firms like J. P. Morgan began making it a practice to hire the children of Chinese government officials to curry favor with their powerful parents. These children are sometimes called "Princelings" because they are the children of senior government officials. J. P. Morgan was ultimately charged with violating the Foreign Corrupt Practices Act (FCPA) by U.S. officials for engaging in the "Princeling" practice. Whether they should have been charged or not is a question for another time.[36] But it is important to note that the FCPA prevents *American* corporations from hiring or doing special business deals with the children of *foreign officials*. It does *not* prevent foreign entities from hiring or doing special deals with the *children of American officials*.

Whatever the case, the fact remains that those prospering the most in China are often the relatives of powerful officials, especially their children. The Princelings have become a powerful source of deal making in a country where familial ties matter. Having the proper connections or family ties in China is so important that there is a single word for it: *guanxi* (pronounced gwan-Chē). It is a word that describes the system of social networks that facilitate business in China. The word means more than its literal English translation of "connection" or "relationship" because those words do not capture the power of mutual obligation, reputation, and trust that are central to *guanxi*. Having *guanxi* refers to a certain weightiness of the people involved and with whom they are ultimately associated.[37]

While *guanxi* in business and personal matters may be a noble cultural practice, when it involves American politicians and their families, whose weightiness comes from their elected office, *guanxi* crosses the line into corruption.

Such was the case when Rosemont Seneca joined forces in doing business in China with another politically connected consultancy called the Thornton Group.[38] The Massachusetts-based firm is headed by James Bulger, the nephew of the notorious mob hit man James "Whitey" Bulger. Whitey was the leader of the Winter Hill Gang, part of the South Boston mafia. Under indictment for nineteen murders, he disappeared. He was later arrested, tried, and convicted.[39]

James Bulger's father, Whitey's younger brother Billy Bulger, serves on the board of directors of the Thornton Group.[40] He was the longtime leader of the Massachusetts State Senate and, with their long overlap by state and by party, a political ally of Massachusetts senator John Kerry.[41]

Less than a year after opening Rosemont Seneca's doors, Hunter Biden and Devon Archer were in China having secured access at the highest levels. Thornton Group's account of the meeting on their Chinese-language website is telling: Chinese executives "extended their warm welcome" to the "Thornton Group, with its US partner Rosemont Seneca chairman Hunter Biden (second son of the now Vice President Joe Biden)." The purpose of the meetings was to "explore the possibility of commercial cooperation and opportunity."[42] Curiously, details about the meeting do not appear on their English-language website.

Also, according to the Thornton Group, the three Americans met with the largest and most powerful government fund leaders in China—even though Rosemont was both new and small. To put these meetings in perspective, it was as if the son of the Chinese premier held a single meeting with the heads of Goldman Sachs, Bank of America, J. P. Morgan, Merrill Lynch, and Blackstone. Except, in this case, these were government entities with trillions of dollars of capital to invest. The delegates spent two days meeting with the top executives from China's sovereign wealth fund, social security fund, and largest banks. Hunter posed with them for a series of pictures.[43]

The timing of this meeting was also curious. It occurred just hours before Hunter Biden's father, the vice president, met with Chinese president Hu in Washington as part of the Nuclear Security Summit.[44]

There was a second known meeting with many of the same Chinese financial titans in Taiwan in May 2011.[45] For a small firm like Rosemont Seneca with no track record, it was an impressive level of access to China's largest financial players. And it was just two weeks after Joe Biden had opened up the U.S.-China strategic dialogue with Chinese officials in Washington.[46]

On one of the first days of December 2013, Hunter Biden was jetting across the vast expanse of the Pacific Ocean aboard Air Force Two with his father and daughter Finnegan.[47] The vice president was heading to Asia on an extended official trip. Tensions in the region were on the rise.

The American delegation was visiting Japan, China, and South Korea. But it was the visit to China that had the most potential to generate conflict and controversy. The Obama administration had instituted the "Asia Pivot" in its international strategy, shifting attention away from Europe and toward Asia, where China was flexing its muscles.[48] Before the plane took off, Beijing had unilaterally declared a new air defense zone over disputed territories in the East China Sea. Other countries, notably Japan and South Korea, also laid claim to the territory.[49] The area was reportedly rich in energy and natural resources, making it a valuable economic prize.[50] The Chinese move was a declaration of sovereignty requiring international airlines to register their flights with the Chinese military before flying over the disputed territory.[51] Biden's visit was under scrutiny across Asia.

For Hunter Biden, the trip coincided with a major deal that Rosemont Seneca was striking with the state-owned Bank of China. From his perspective, the timing couldn't have been better.

Before China, they briefly stopped in Japan. The vice president held a series of meetings with the Japanese deputy prime minister and members of the Japanese parliament. Later he met with the Japanese crown prince and finally dined with Japanese prime minister Abe. Japan was concerned about the Chinese move, but

Biden said little on the subject publicly. What he did say did not placate Japan. He voiced concern about China's air defense zone but did not call for a reversal.[52] On Wednesday, December 4, the delegation flew on to Beijing where the heavy lifting was expected to take place.

Vice President Biden, Hunter Biden, and Finnegan arrived to a red carpet and a delegation of Chinese officials. Greeted by Chinese children carrying flowers, the delegation was then whisked to a meeting with Vice President Li Yuanchao, and talks with President Xi Jinping.[53] As he chatted with Vice President Li, Biden explained how he saw the world: the Beijing-Washington axis was the "central, sort of, organizing principle" of international relations. It was a view that was warmly and eagerly welcomed by Beijing.[54]

Biden spent five and half hours in conversation with President Xi. They "covered every single topic in the U.S-China relationship," as one senior administration official described it.[55]

Despite the recent Chinese action of declaring a defense zone in the East China Sea, the subject did not dominate the talks. Indeed, Biden never even publicly mentioned the defense zone during his visit.[56] This trip was all about "practical cooperation."[57] Biden was hoping, in the words of White House officials, "to build a different kind of relationship for the 21st century."[58] As one regional observer put it:

> A somber and tired Biden couldn't break the ice except by urging Beijing to exercise caution and restraint in policing the zone to avoid accidents and miscalculations. In the end, the United States allowed its commercial carriers to comply with China's requirements for flight information, to the dismay and disappointment of Japan.[59]

That night, the Bidens stayed at the luxurious St. Regis Hotel, and the following morning they met with the U.S.-China Business Council.[60] Then it was off to Villa No. 5 of the Diaoyutai State Guesthouse. This was where Richard Nixon stayed during his famous 1972 visit, and where Madame Mao made permanent residence during the Cultural Revolution. Biden, wearing a light blue tie and dark suit, had another meeting with Vice President Li. When Biden arrived, they "chatted as if they were old chums," according to the media pool reporter. Vice President Li Yuanchao told pool reporter Steve Clemons of the *Atlantic*, "Vice President Biden is a good friend to you and to me."[61]

Hunter and Finnegan Biden joined the vice president for tea with U.S. ambassador Gary Locke at the Liu Xian Guan Teahouse in the Dongcheng District in Beijing.[62] Where Hunter Biden spent the rest of his time on the trip remains largely a mystery. There are actually more reports of his daughter Finnegan's activities than his.

One of the few public sightings of Hunter Biden occurred after that tea, when Vice President Biden and his son, along with granddaughter Finnegan, halted the motorcade and began walking along the south side of a shopping street. Hunter Biden, dressed in a dark overcoat covering a deep blue zip-up sweater, tagged along as his father stepped into a small shop. With the world's media in tow, they emerged a few minutes later with the vice president holding a Magnum ice cream bar.[63]

The vice president raised his Magnum ice cream bar to show the world's press how personable Joe Biden could be, defense zone aside. Intentionally or not, Hunter Biden was showing the Chinese that he had *guanxi*.

The picturesque visit to the shop and the deep diplomatic significance of the visit were widely reported by the Chinese and world media. What was not reported was the deal that Hunter was securing. Rosemont Seneca Partners had been negotiating an exclusive deal with Chinese officials, which they signed approximately ten days after Hunter visited China with his father.[64] The most powerful financial institution in China, the government's Bank of China, was setting up a joint venture with Rosemont Seneca.

The Bank of China is an enormously powerful financial institution. But the Bank of China is very different from the Bank of America. While both have massive financial resources, the similarity ends there. The Bank of China is government owned, which means that its role as a bank blurs into its role as a tool of the government. The Bank of China provides capital for "China's economic statecraft," as scholar James Reilly puts it. Bank loans and deals often occur within the context of a government goal.[65]

As one Chinese scholar puts it, these government-owned entities are difficult for Westerners to fully comprehend because the methods of government control are deep. "Chinese SOEs [state-owned enterprises] are embedded in a complex state-controlled network" including state ministries, security forces, and the Communist Party.[66] As a result, such a bank is extraordinarily linked to the government's ruling elite. For example, the chairman of the Bank of China, Tian Guoli, also serves as secretary of the Communist Party of China Party Committee at the bank.[67] Tian was familiar with both Joe Biden and John Kerry. Vice President Biden had met with executives of the Bank of China during his 2011 official visit to China.[68]

Ten days after Hunter Biden left China with his father, Rosemont Seneca and the Bank of China created an investment fund called Bohai Harvest RST (BHR), a name that reflected who was involved. Bohai (or Bo Hai), the innermost gulf of the Yellow Sea, was a reference to the Chinese stake in the company.[69]

The "RS" referred to Rosemont Seneca. The "T" was Thornton.[70] The fund enjoyed an unusual and special status in China. BHR touted its "unique Sino-U.S. shareholding structure" and "the global resources and network" that allowed it to secure investment "opportunities."[71] Funds were backed by the Chinese government.[72]

In short, the Chinese government was literally funding a business that it co-owned along with the sons of two of America's most powerful decision makers. The Chinese government-funded Bohai Harvest RST prominently mentioned Rosemont Seneca's involvement in the fund on its website, that is, until inquiries were made about Biden's and Heinz's involvement. Subsequently, any mention was scrubbed or abbreviated "RST."

Jonathan Li, chief executive officer of Bohai Harvest, says, "BHR represents a mixture of private enterprise and SOE [state-owned enterprise]." Xin Wang, managing partner of the firm, notes the advantages of having a connected Western business partner like Rosemont Seneca. "Just by virtue of being an SOE there is the perception—rightly or wrongly [*sic*]—that there will be some cross-cultural issues. Having us and our global resources there as a financial investor, and serving as a conduit, can facilitate the transaction."[73]

Rosemont Seneca's role, according to internal Chinese documents, was to spearhead efforts in the United States.[74]

The deal was remarkable. Rosemont Seneca was getting something for the first time that no other Western firm had in China, a private equity cross-border investment fund formed in the Chinese government's Shanghai Free-Trade Zone.[75] The Shanghai Free-Trade Zone had been established only months earlier by the Chinese government's State Council, and "personally championed" by China's prime minister, Li Keqiang.[76] Why, with all the financial firms doing business in China that had far more experience, a larger footprint, and a history of doing deals there, did the Chinese government pick Rosemont Seneca?

The advantages that Biden and Heinz's firm got were enormous. Along with Chinese government capital to invest, by operating from the Shanghai Free-Trade Zone they could take Chinese government funds and invest there, or take them out of the country and invest them in the United States or elsewhere. No one else had such an arrangement in China.

Rosemont Seneca was essentially placed first in line.

The Chinese government's deal in December 2013 with the sons of America's vice president and secretary of state occurred in lockstep with aggressive territorial claims it was making in the Pacific. Ely Ratner, writing in *Foreign Affairs*, described how "in early 2014, China's efforts to assert authority over the South China Sea went from a trot to a gallop. Chinese ships began massive dredging projects to reclaim land around seven reefs that China already controlled in the Spratly Islands, an archipelago in the sea's southern half. In an 18-month period, China reclaimed nearly 3,000 acres of land." Contrary to assurances by the Chinese president Xi Jinping that they had "no intention to militarize the South China Sea," Beijing began "rapidly transforming its artificial islands into advanced military bases, replete with airfields, runways, ports, and antiaircraft and antimissile systems. In short order, China has laid the foundation for control of the South China Sea."[77]

The Rosemont Seneca deal was lucrative for the Bidens and Kerry's stepson, and appears, on behalf of the Chinese government, strategically timed to go along with China's aggressive territorial claims. But it was not the only deal that occurred at this critical time.

The following year, in July 2014, Secretary of State John Kerry arrived in Beijing for a series of sensitive, high-level meetings with Chinese government officials. As the *Washington Post* put it, Chinese and U.S. relations were in a "downward spiral."[78] Kerry was there as part of the so-called Strategic and Economic Dialogue (S&ED), which had been erected to provide a forum to hash out such tough diplomatic, commercial, and military issues between the two rivals. In 2014, there were several of those issues: sanctions against North Korea; opening up China to greater investment from the West; containing and dealing with Iran's nuclear program; those Chinese territorial claims in the South China Sea, and a climate change agreement.[79]

Before the talks began, Kerry, dressed in a blue blazer and khakis, walked the Great Wall of China. He was joined by the U.S. ambassador to China Max Baucus and Treasury secretary Jack Lew.[80] He was cautiously optimistic about what might be accomplished.

The Chinese government had greeted his appointment as secretary of state a year earlier with a sense of relief. They found his predecessor, Hillary Clinton, too hawkish on China for their tastes. As a lengthy editorial in the official government *China Daily* put it, "Clinton always spoke with a unipolar voice and never appeared interested in the answers she got. Kerry understands the true multipolar nature of the 21st century world. He listens to the answers he gets." They liked his low-key and quiet style. "Although he may not abandon traditional US concerns on promotion of democracy and human rights issues, he will express them in talks privately and quietly, without trying to embarrass or undermine his interlocutors." With his arrival at Foggy Bottom, they surmised, "there is . . . an excellent chance of Sino-US ties improving significantly during his term of office."[81]

Even more auspiciously, the deal between his stepson's Rosemont Seneca and the Bank of China had been inked seven months earlier.

Kerry spent two days in intensive talks with Chinese officials in Beijing behind closed doors. The morning of July 9 commenced with John Kerry standing next to Chinese president Xi Jinping in front of a collection of interspersed Chinese and American flags. President Xi called for a commitment on both sides to "boost Sino-US economic ties." Kerry echoed these sentiments before the leaders went into a private session.[82] Much of what was discussed occurred behind closed doors, but there were some flashes for the public. After a lunch in the Great Hall of the People, John Kerry was invited to pick up a musician's guitar, which he strummed for the assembled guests.[83]

In private discussions, Kerry's focus was on North Korea and Chinese hacking of American computers.[84] Publicly he touted the economic ties between the two countries. As he put it, "China and the United States represent the greatest economic alliance trading partnership in the history of humankind, and it is only going to grow."[85]

Meanwhile, as the secretary of state was engaged in high-stakes secret discussions with his Chinese counterparts in Beijing, other private, commercial discussions were being held with Chinese officials.[86] These talks did not involve the U.S. government. A former subsidiary of a Chinese government company that was close to the Chinese military was beginning discussions with *another* Rosemont entity. The Chinese company, Gemini Investments, was interested in purchasing Rosemont Realty, a firm controlled by John Kerry's stepson and whose leadership included Hunter Biden and close Kerry adviser Devon Archer.[87] At this important diplomatic juncture in U.S.-Chinese relations, with so much at stake in military, diplomatic, and economic terms, a Chinese government–connected company was looking to become business partners yet again with the families of America's secretary of state and vice president. The eye-popping, multibillion-dollar deal that would result a year later would make those connected to the company a lot of money and raise serious questions about a massive conflict of interest.

Archer and others were all too happy to tout Rosemont Realty's political ties to Vice President Joe Biden. As Archer was described in a 2014 corporate biography, "Mr. Archer is the General Partner and Chairman of Rosemont Realty which he founded together with Mr. Hunter Biden."[88] Rosemont Realty played up the connection when pitching opportunities for investors. In a company prospectus from Rosemont Realty watermarked "CONFIDENTIAL," it was a "key consideration" that "Hunter Biden (son of Vice President Biden) is on the advisory board."[89]

To run Rosemont Realty, Heinz, Archer, and Biden tapped another close Kerry confidant to run the day-to-day operations. They tasked fellow Yale graduate Daniel Burrell with building the real estate business from the ground up. Burrell, who hails originally from upstate New York, worked on Kerry's 2004 failed presidential campaign. Burrell called John Kerry "an incredible mentor."[90] A *Los Angeles Times* article from October 2004 described Burrell's move to Los Angeles to run the local campaign. He came to L.A. with "no car, no place to stay," living, as a favor to Kerry, at one of his supporters' place, and arranged for fundraisers featuring Hollywood types.[91]

In 2009, Burrell ventured out to the foothills of north-central New Mexico to set up Rosemont Realty's corporate headquarters. He settled into a building in the state capital, Santa Fe, on historic Garfield Street near the Railyard, a cultural hub in the center of the city. Rosemont put the mayor of Santa Fe on the payroll, which is legal in New Mexico.[92] Within a year, the company had purchased the state's largest commercial real estate company, BGK, which had a $1.5 billion commercial real estate portfolio. As Burrell describes it, he was soon "jetting around looking for trophy buildings" in several states.[93] Within a matter of the next four years, the company accumulated office buildings from the Southwest to New England. Rosemont Realty erected regional offices outside their headquarters: in Albuquerque; Atlanta; Dallas; Denver; Houston; New York; Peoria, Arizona; San Antonio; Washington, D.C.; and Tulsa.[94]

In mid-2014, just as the Rosemont Seneca deal with the Bank of China was getting off the ground, Burrell began holding discussions with a Chinese subsidiary called Gemini Investments Limited about buying Rosemont Realty.[95] Gemini might have appeared to be just like any other investment company; publicly traded on the Hong Kong Stock Exchange, it offered itself as a conventional real estate investor. However, as can often be the case with China, tracing corporate ownership yields some interesting results. While Gemini Investments is a publicly traded stock, it is called an "indirect subsidiary," and control of the company is retained by its parent company, Sino-Ocean Land. Gemini Investments' director and honorary chairman is Li Ming. His corporate ties are to some of the most sensitive and politically important corporations in China. He also served for several terms running as a member of the Chinese Communist Party's elite so-called People's Political Consultative Conference.[96]

Gemini Investments' parent, Sino-Ocean Land, is a development company with a "large footprint in Beijing," as one journalist put it, and is "one of the largest real estate companies in Beijing."[97] In an old bio, Devon Archer describes it as "the largest

SOE real estate developer in China."[98] It has deep ties to the Chinese government. Sino-Ocean Land, for example, launches and funds programs under the direction of the Central Commission of the China Communist Youth League and the government's Ministry of Education, according to corporate records.[99]

In turn, Sino-Ocean Land is connected to a company called the China Ocean Shipping Company (COSCO). The chairman of Sino-Ocean Land is also chairman of COSCO.[100] The Sino-Ocean Land website states that "Sino-Ocean Group originated from COSCO." Until 2011, Sino-Ocean Land was a subsidiary of COSCO, and one of its website backgrounds is a photo of the COSCO headquarters; they share the same address.[101]

COSCO was founded in 1961 and is a state-owned company.[102] With palatial headquarters in Ocean Plaza in the Xicheng District in Beijing, its corporate offices face the headquarters of another pillar of Chinese government power, the Bank of China—Rosemont's partner in the Bohai Harvest deal.[103] COSCO is one of those state-owned enterprises (SOEs) that populate the corporate world in China. But even within the ranks of Chinese government–controlled companies, COSCO is unique. COSCO is widely seen by authorities in Asia and the West as deeply embedded in and tied to the Chinese military establishment. In particular, COSCO had strong organizational links with both the Chinese People's Liberation Army (PLA) and the People's Liberation Army Navy (PLAN). As the academic journal *Maritime Affairs* puts it, "It is well known that the state-owned China Ocean Shipping Company (COSCO) has close links with the PLA. The Chinese call COSCO the fifth arm of the PLAN, and [COSCO's] ships are often referred to as 'zhanjian' (warships)."[104] COSCO has a long history of clandestine work overseas and general troublemaking as far as American interests are concerned. According to leaked secret State Department cables, COSCO was reportedly shipping material for Syria's weapons development program—probably from North Korea.[105] In 2015, officials in Colombia detained a COSCO vessel that was alleged to have been illegally shipping thousands of cannon shells and explosive material. (The company had listed the cargo as "grain" on a shipping manifest.)[106] In 1996, two thousand smuggled Chinese assault rifles were seized near San Francisco, and were known to have been brought through the Port of Oakland on a COSCO ship.[107]

More broadly, COSCO has built a global footprint covering vast corners of the world. U.S. naval experts say that those commercial activities by COSCO are perfectly structured to align with Chinese naval strategy. According to Rear Admiral Michael McDevitt, USN (retired), "With the China Ocean Shipping Company, a state-owned enterprise providing global logistical services, Beijing enjoys built-in shore-based support structure at virtually all the major ports along the Pacific and Indian Oceans . . . this has become a successful approach to logistic sustainment halfway around the world from Chinese bases."[108]

So when you navigate your way through the ownership of the Rosemont entities in the United States and Gemini Investments, it becomes strangely clear that a company connected to sons of the vice president and secretary of state was negotiating to secure a deal with a company whose ties could be traced back to the "fifth arm" of the Chinese navy.

It would be the second large and profitable deal that the son of the vice president and the stepson and friends of John Kerry would strike with Chinese government–connected companies as both statesmen were negotiating with Beijing.

In short, as Secretary of State John Kerry and Vice President

Joe Biden were engaged in sensitive, high-stakes negotiations with the Chinese government, their sons' companies were cutting yet another deal with a company connected to the Chinese government.

During his tenure as secretary of state, Kerry was criticized for being soft on China despite China aggressively laying claim and expanding its presence in the South China Sea. Alarm bells were going off all over the region because of this unilateral expansion. But Kerry played it cool. As *Bloomberg News* put it in one headline, "Kerry's Soft Words Blunt U.S. Hard Power in the South China Sea."[109]

Critics noted that when it came to Chinese territorial claims in Asia, Beijing wanted to have negotiations with countries in the region individually, excluding both the United States and Japan, to make it easier for the large power to intimidate smaller regional players who questioned their territorial claims. In a move that surprised and troubled some in the region, Kerry effectively endorsed China's strategy to isolate countries like the Philippines in these negotiations by refusing to have the United States take a side in the territorial dispute.[110] As another observer put it, "If these efforts by China succeed, then the U.S. (and Japan) will effectively be sidelined. Which itself would be a win for the Middle Kingdom."[111]

Kerry was also clear to state publicly that he saw no need to "contain" China. This was in contrast to his predecessor Hillary Clinton's posture. His words and position were praised by Beijing.[112]

Chinese government officials praised Kerry for his low-key approach to relations with China. Consider this dispatch from the Chinese embassy about a meeting between Kerry and Chinese foreign minister Wang Yi: "John Kerry agreed with the ideas and way of thinking from the Chinese side on developing the US-

China relations. He said that the US side appreciates China for its positive and conducive efforts in promoting the talks on [the] Iranian nuclear issue and the political settlement of the Syrian issue and others. The two sides should continue to strengthen communications and consultations in major regional hot spot issues, and to enrich the contents of the new model of major-country relationship between the US and China."[113]

Privately, the business negotiations between the Biden and Kerry families and Chinese entities continued. Publicly, Secretary of State Kerry engaged with the very same Chinese government in diplomatic negotiations. In November 2014, Kerry hosted the Chinese foreign minister in his hometown of Boston, where they dined on Maine cod and Boston cream pie while overlooking Boston Harbor.[114]

By December 2014, Gemini was negotiating and sealing deals with Chris Heinz's and Hunter Biden's Rosemont on several fronts. That month, Gemini bought out the Rosemont Opportunities Fund II, an offshore investment vehicle run by Rosemont, for $34 million. Larger deals would follow.[115]

In May 2015, Kerry went to Asia to meet with his Chinese counterparts to readdress the difficult issues between the United States and China. As Josh Rogin recounts, "Chinese officials told me that the Chinese saw further weakness when Secretary of State John Kerry visited Beijing in May. According to the Chinese readouts, Kerry told his hosts that the U.S. wanted to work with them on a range of issues, including North Korea, Iran and Syria, and the two powers shouldn't let the South China Sea issue get in the way of broader cooperation. The Chinese interpreted it as a signal that the U.S. was not ready to confront them."[116]

By August, Rosemont Realty announced that Gemini Investments, still run from COSCO headquarters, was buying a

75 percent stake in the company.[117] It was the second major deal Rosemont struck with China in just eighteen months. The terms of the deal included a $3 billion commitment from the Chinese, who were eager to purchase new U.S. properties.[118]

Rosemont Realty was rechristened Gemini Rosemont.

"Rosemont, with its comprehensive real estate platform and superior performance history, was precisely the investment opportunity Gemini Investments was looking for in order to invest in the U.S. real estate market," declared Li Ming, Sino-Ocean Land Holdings Limited and Gemini Investments chairman. "We look forward to a strong and successful partnership."[119]

The plan was to use Chinese money to acquire more properties in the United States. "We see great opportunities to continue acquiring high-quality real estate in the U.S. market," one company executive said. "The possibilities for this venture are tremendous."[120]

So during a critical eighteen-month period of diplomatic negotiations between Washington and Beijing, the Biden and Kerry families and friends pocketed major cash from companies connected to the Chinese government.

The consequences of those deals are as surprising as the fact that they were conducted in the first place.

3

NUCLEAR AND OTHER CONSEQUENCES

- The circle of family and confidants around Biden and Kerry participated in private deals that ended up serving Chinese military and strategic interests to the detriment of the United States.
- A Chinese company in which they invested was stealing U.S. nuclear secrets, according to the FBI.

The partnership between American Princelings and the Chinese government was just a beginning. The actual investment deals that this partnership made were even more problematic.

The initial announcement of Rosemont Seneca's and the Chinese government's Bohai Harvest joint-investment fund in December 2013 revealed plans for a $1 billion investment. Seven months later, they increased it to $1.5 billion.[1] The deals began to emerge almost immediately. Many of them would have serious national security implications for the United States.

Bohai Harvest RST (BHR), with funds provided by the Chinese government, not surprisingly gained access to private-equity deals and highly prized initial public offerings (IPOs) affected by the strategic "privatization" of Chinese state-owned firms. Many were listed on the Hong Kong stock exchange.

BHR investments generally fit a pattern of buying stakes in

companies that controlled technology of interest to the Chinese government. When BHR took Chinese funds and invested in the United States, it bought up companies that fit the same profile.

It is important to keep in mind that Rosemont Seneca was not passively involved in the investment decisions being made. Devon Archer, close Kerry confidant who helped run the Heinz Family Office, served as the vice-chairman of BHR with responsibility for investments. As he has explained on his website, "As vice chairman of Bohai Harvest RST . . . Mr. Archer helps guide one of China's most prominent mutual fund management companies."[2]

In 2015, BHR joined forces with the automotive subsidiary of the Chinese state-owned military aviation contractor Aviation Industry Corporation of China (AVIC) to buy American "dual-use" parts manufacturer Henniges.[3] Formed initially during the Korean War as the Bureau of Aviation Industry, AVIC is a major military contractor in China.[4] AVIC owns 50 percent of the military arms business of CATIC—the China National Aero-Technology Import and Export Corporation.[5] It operates "under the direct control of the State Council" and produces a wide array of fighter and bomber aircraft, transports, and drones—primarily designed to compete with the United States.[6] The company also has a long history of stealing Western technology and applying it to military systems. The year before BHR joined with AVIC, the *Wall Street Journal* reported that the aviation company had stolen technologies related to the U.S. F-35 stealth fighter and incorporated them in their own stealth fighter, the J-31.[7] AVIC has also been accused of stealing U.S. drone systems and using them to produce their own.[8] In 2011, AVIC and the People's Liberation Army signed an "Agreement on Military and Civilian Integrated Support for New Army Aviation Equipment."[9]

In short, AVIC sits at the heart of the Chinese military industrial complex.

In September 2015, when AVIC bought 51 percent of American precision parts manufacturer Henniges, the other 49 percent was purchased by the Biden-and-Kerry-linked BHR. Henniges is recognized as a world leader in anti-vibration technologies in the automotive industry and for its precise, state-of-the-art manufacturing capabilities.[10] Anti-vibration technologies are considered "dual-use" because they can have a military application, according to both the State Department and Department of Commerce.[11] Part of Henniges had previously been owned by GenCorp, an American rocket and missile manufacturer now known as Aerojet Rocketdyne Holdings, Inc.[12] The technology is also on the restricted Commerce Control List used by the federal government to limit the exports of certain technologies.[13] For that reason, the Henniges deal would require the approval of the Committee on Foreign Investment in the United States (CFIUS), which reviews sensitive business transactions that may have a national security implication.[14] The U.S. intelligence community and congressional intelligence committees have long recognized that the Chinese government is "committed to the acquisition of Western machine tool technology," especially those where it is difficult for outsiders "to distinguish between civilian and military end-uses of the equipment."[15]

According to BHR internal documents, the Henniges deal included "arduous and often-times challenging negotiations."[16] The CFIUS review in 2015 included representatives from numerous government agencies including John Kerry's State Department.[17] The deal was approved in 2015.[18]

Then there was the involvement of Rosemont Seneca in the investment in a controversial nuclear power company. In December

2014, the Biden-and-Kerry-linked BHR became, in Bohai's words, an "anchor investor" in China General Nuclear Power Corporation (CGNPC or CGN), a state-owned nuclear company involved in the development of nuclear reactors.[19] CGN had been wholly owned by the Chinese government and was now selling off a small stake to outside investors. Because the nuclear industry is considered a strategic sector in China, it would remain under the direct control of the Chinese State Council, and the government would remain the controlling shareholder.[20]

BHR also got a piece of Sichuan Sanzhou Special Steel Pipe (SZSSP), which manufactures special piping for use in the nuclear and petrochemical industries, including cooling systems for nuclear reactors.[21]

That Kerry allies and Biden family members were making money courtesy of the Chinese government was troubling enough. The quagmire of Rosemont's participation in these sensitive Chinese industries deepens with the additional fact that CGN was under FBI investigation and eventually charged with stealing U.S. nuclear secrets.

In November 2013, a Chinese American engineer traveled to China at the expense of the Chinese government. Ching Ning Guey was paid by the Chinese government for sensitive information on American nuclear technology. He worked for the Tennessee Valley Authority (TVA) and passed along tightly controlled information concerning American nuclear reactors, which was then being transferred to CGN.

Guey soon received the attention of the FBI. (He eventually pleaded guilty to federal charges and would cooperate with the FBI in its broader investigation of CGN.)[22] Guey started talking, and soon led the FBI to another engineer named Allen Ho, who, ironically, lived in Wilmington, Delaware, just five minutes from

Joe Biden's house.[23] Allen Ho, under the direction of CGN, had been working to steal American nuclear secrets since 1997. According to the FBI, he was looking for information that would help CGN's small modular reactor program, advanced fuel assembly programs, fixed in-core detector system, and "verification and validation of nuclear reactor–related computer codes." These were prized secrets that engineers were not allowed to share or discuss with foreign engineers, companies, or government officials.

In April 2016, Ho and CGN were charged by the U.S. Justice Department with stealing nuclear secrets from the United States—actions prosecutors said could cause "significant damage to our national security."[24]

This was no haphazard spy ring. The federal government's U.S.-China Economic and Security Review Commission cited the Ho case in its annual report as an example of Chinese espionage that constituted a "large and growing threat" to American national security.[25]

According to U.S. attorneys, CGN instructed Ho to hire American nuclear engineers as consultants and attempt to get them to share sensitive information. Other information was obtained by outright theft.[26] Of particular interest to CGN was Westinghouse's AP1000 system, an advanced nuclear reactor that has military application.[27] It could be used commercially to generate power for civilians. But it also had application for developing nuclear submarines.

When Ho was arrested by the FBI he was using a "random code generator" so he could gain access to funds provided to him via the Bank of China.[28]

Why the Chinese interest in the Westinghouse AP1000? Even though the AP1000 is a civilian nuclear reactor, the design poses a "dual-use" concern, according to American national security officials.[29] In other words, it could have important military applications, especially in submarines. Piping systems that were a particular target of this espionage strongly resemble the coolant pumps used in U.S. nuclear submarines.[30]

CGN's espionage efforts were intense and prolonged. According to the indictment, Ho allegedly, under the direction of CGN executives, "also identified, recruited, and executed contracts with U.S.-based experts from the civil nuclear industry who provided technical assistance related to the development and production of special nuclear material of CGNPC in China. Ho and CGNPC also allegedly facilitated the travel to China and payments to the U.S.-based experts in exchange for their services."[31]

The Obama Justice Department was blunt: Ho "conspired with others to knowingly act as an agent of China." He told recruits that "China has the budget to spend." According to DOJ documents, "Ho made clear that he was charged with obtaining necessary expertise from the United States at the direction of the CGNPC and the China Nuclear Power Technology Research Institute, a subsidiary of CGNPC, and that he was to do so surreptitiously."[32]

It is clear from the transcripts of the FBI's interrogation of Ho that they considered the stakes in the investigation high and that what they ultimately prized was intelligence on CGN. There is no evidence that the FBI was aware at the time that Biden and Kerry family members or confidants were involved with the company.

During the interrogation, FBI agents told Ho: "You give us a deep understanding of how things at CGNPC, CNPRI, how they go up, how they are—how the government of China pushes down, and how they interact . . . We think that you have knowledge. You've been over there a long time. You're a senior consultant [to CGN], an executive consultant. I've seen pictures—I've seen emails where you talk about the CEO being your friend."[33]

The FBI was also cognizant of the fact that because CGN was a state-owned company it was a diplomatically sensitive case, but because of its serious nature, they were intent on pursuing it. As one agent told Ho during the interrogation, "It's a big statement for the U.S. government to charge a Chinese state-owned company, and they don't take it lightly."[34]

What the Chinese government couldn't buy from the United States, it would steal. Chinese companies have a long history of stealing American nuclear secrets. As scholar Ralph Sawyer puts it in his masterful book *The Tao of Spycraft*, "No nation has practiced the craft of intelligence or theorized about it more extensively than China."[35]

This time, of course, family members of America's most powerful political figures had exclusive financial interests in the rogue Chinese company stealing American secrets.

Anyone with a basic knowledge of Chinese commercial espionage, particularly in a sector like nuclear power, should know that spying will play a role in the company's activities. The involvement of Kerry confidants and Biden's son in CGN should have set off alarm bells for all of them.

Repeated attempts to seek comments from Joe Biden, John Kerry, Hunter Biden, Chris Heinz, and Devon Archer have gone unanswered.

When did the partners in Rosemont learn of CGN's intelligence activities? What we do know is that after the charges were filed and Ho pleaded guilty, Rosemont did not change its relationship with its Chinese partners, nor did BHR divest from the state-owned Chinese company that had been stealing America's nuclear secrets.

In 2016, the Biden-and-Kerry-linked BHR made another strategic investment in unison with a Chinese state-owned company, China Molybdenum. Over the past decade, China had been in an intense race to accumulate mineral assets from around the world. China Molybdenum operates to accumulate its namesake and other rare-earth minerals. Molybdenum is used to strengthen metals, particularly to make "tough, steel alloys," and is found "as a by-product of copper mining."

During World War I, when the British introduced the tank, they made an important discovery. One inch of steel fused with molybdenum offered better protection against a direct hit than three inches of regular (manganese) steel. It was soon being used on a whole variety of military vehicles and applications.[36] It is also used in the nuclear industry.[37]

The company, China Molybdenum, does not hide the fact that the minerals it acquires have direct military application. In its 2015 annual report, the company explains how its products are used: "Molybdenum alloy steel is used for manufacturing alloy components and parts of warships, tanks, guns and cannons, rockets and satellites." The company is also a major producer of tungsten, which, as it points out, "is widely used in military engineering."[38]

The company, as you might expect, has deep ties to the Chinese government and the Communist Party of China.[39]

China Molybdenum is China's "biggest producer of molybdenum" and China and the United States have been locked in a global battle for the control of molybdenum assets. As one U.S. mining executive described it in the magazine *National Defense*, "Molybdenum, for example, is a key component in the manufacturing of armor plating . . . Unfortunately, it's increasingly becoming a zero-sum game to procure these minerals as demand soars and the United States' access to these resources is put in jeopardy."[40]

Just how intense is the battle between China and the United States for control of critical minerals?

In 2012, President Obama entered the Rose Garden of the White House to announce that the Obama-Biden administration was filing a complaint against China with the World Trade Organization.[41] The administration was charging Beijing with attempting to control the world rare-earth minerals market, including restricting Western access to molybdenum. Japan and the European Union joined the Obama administration in the complaint, and in March 2014 the WTO ruled that China was indeed violating international agreements in what the *New York Times* called a "hard" ruling against Beijing.[42]

How ironic then that barely two years later, on May 9, 2016, China Molybdenum announced that it was buying more than half of the massive Tenke Fungurume copper mine in the Democratic Republic of Congo for $2.65 billion. Tenke is regarded as "one of the world's prized copper assets."[43] Even more ironic: a few months later, BHR announced that it was buying another 24 percent stake in the mine for $1.5 billion.[44] China Molybdenum helped BHR acquire its ownership stake, thereby further consolidating Chinese control over the coveted mine.[45] The mine is also an important source of cobalt, a critical resource for China's lithium-ion battery industry.[46]

So while the Obama administration was laying out the challenge posed by China in the global minerals race, the son of the vice president and a confidant of the secretary of state were invested in deals that would help Beijing win that resource race.

In short, these investments served the Biden-Kerry families' investment interests, and also served Chinese strategic interests. On top of it all, these deals served ultimate Chinese strategic interests by forging financial ties with the families of key American politicians. They provided a legitimizing factor, a Trojan horse, if you will, for these Chinese investments and made them seem far more benign than they actually were.

Vice President Joe Biden and Secretary of State John Kerry were managing policy on America's relationship with China while their sons and close associates were receiving deals from the Chinese government and government-related entities. The next chapter shows how the business model was not limited to China.

4

BIDENS IN UKRAINE

- While Vice President Biden was overseeing U.S. policy toward Ukraine, his son joined the board of one of Ukraine's most profitable and corrupt energy companies.
- The Bidens could be potential billionaires as a result of the deal.
- Did Joe Biden look the other way when $1.8 billion in taxpayer money disappeared?

Half a world away, the father-statesmen, Biden and Kerry, were also deeply involved in a complex and difficult situation in Ukraine. The country was facing Russian aggression in the east, political turbulence at home, and powerful oligarchs fighting over the country's resources. In the middle of it Kerry and Biden were making critical decisions about Ukraine's future with Hunter Biden and Devon Archer on their heels.

Ukraine is blessed with a treasure house of energy resources, particularly oil and natural gas. These riches are a target for Russian covert operations, corrupt oligarchs, and powerful mafia dons, who all sought to control them. Burisma is a secretive Ukrainian natural gas company with deep political ties in the country, and the nation's second-largest private natural gas producer. Many of Burisma's assets are heavily concentrated in the contentious eastern reaches of the country. This border area with Russia has been the source of conflict with Moscow for decades.

Burisma was created in 2006 with a Cypriot registration by Mykola Zlochevsky, the barrel-chested, bald-headed, and future ecology and natural resources minister under the pro-Russian government of Viktor Yanukovych. How did a government minister end up owning a massive energy company? In a story far too familiar in that part of the world, Zlochevsky gave himself the licenses to develop the abundant gas fields. Of note, he took the license for the country's largest natural gas field, Sakhalinska, from someone else and gave it to a company connected to Burisma.[1]

To complete the portrait, Zlochevsky had a reputation for lavishness while in government service, with a taste for Bentleys and the occasional Rolls-Royce.[2] His other business ventures also reflect his reputed lifestyle. He owns a super-exclusive fashion boutique in downtown Kiev named Zlocci. With chandeliers, marble-topped tables, and recessed interactive panels, Zlocci sells accessories made of alligator, ostrich, eel, and lizard. At Zlocci, a set of matching crocodile dress shoes and belt will set you back $2,800. The average Ukrainian would have to spend nearly nine months' wages to pay for it.[3]

In 2012 President Yanukovych removed Zlochevsky as ecology and natural resources minister and appointed him to the National Security and Defense Council. But fortunately, a year earlier the ownership structure of Burisma had been quietly transferred to a Cyprus-based company called Brociti Investments. The island nation has become a favorite venue for Russian activities, where oligarchs, mafia, government officials, and intelligence operatives park their assets because of the very tight secrecy laws. Zlochevsky's name was still attached to the company, but Burisma's major subsidiaries now listed the same business address as the natural gas firm controlled by a controversial Ukrainian oligarch named Ihor Kolomoisky.[4] The UK's *Guardian* newspaper in 2015 reported Kolomoisky to be perhaps Ukraine's "most troubling oligarch of all."[5] For a country rife with corruption, war, self-dealing, and cronyism, that is saying something. But he would prove to be a worthy business partner for Joe Biden's son and John Kerry's inner circle.

Pudgy with a thick crop of silver hair, wire-frame glasses, and a tight beard, Kolomoisky was born into a family of engineers from the eastern half of Ukraine. His power base was Dnipropetrovsk (since May 2016, Dnipro), an industrial center in the country that has been a cradle to a succession of powerful Ukrainian figures. Dnipro is known as the stomping ground of ex-presidents Leonid Kuchma and Oleksandr Turchynov, as well as oligarchs like Victor Pinchuk. Before the breakup of the Soviet Union, the area was the power base for Soviet leader Leonid Brezhnev. Through his company the Privat Group, Kolomoisky controlled Ukraine's largest financial institution, PrivatBank, through which the Ukrainian military got paid and government pensions were distributed. He also controlled media companies and airlines. Sometimes known as "King Kolomoisky" inside the country, his office features a gigantic shark tank, but he does most of his business from his luxurious home in Switzerland.[6]

Kolomoisky does not seem to care much for the rules. He holds Ukrainian, Israeli, and Cypriot passports, which is a problem because the Ukrainian constitution forbids dual citizenship. When asked about that fact Kolomoisky quipped, "The constitution prohibits double citizenship but triple citizenship is not forbidden."[7]

Kolomoisky's reportedly violent and brutal business practices stand out even in this rough corner of the globe. A British judge upbraided the billionaire for taking control of a company in Ukraine "at gunpoint."[8] Rival oligarchs have sued him in British courts, alleging that he was involved in "murders and beatings" in relation to a prior business deal.[9] Kolomoisky allegedly used "quasi-military teams" and sent "hired rowdies armed with baseball bats, iron bars, gas and rubber bullet pistols and chainsaws" to take over the Kremenchuk Steel Plant in 2006.[10] He also used "a mix of phony court orders (often involving corrupt judges and/or registrars) and strong-arm tactics" to purge rival board members.[11] Kolomoisky vigorously denies any illegal activities.

Why young Biden and those close to Kerry viewed Kolomoisky as an appropriate business partner is not known. Repeated calls and e-mails to both about their work with Burisma went unanswered.

Kolomoisky built his multibillion-dollar empire by "raiding" other companies, a violent Ukrainian form of mergers and acquisitions involving guns. Matthew Rojansky, director of the Kennan Institute at the Woodrow Wilson International Center for Scholars, has done an in-depth study of the practice: "there are actual firms in Ukraine . . . registered with offices and business cards, firms [that specialize in] various dimensions of the corporate raiding process, which includes armed guys to do stuff, forging documents, bribing notaries, bribing judges." Although the practice is common, Kolomoisky stands out. Rojansky calls him "the most famous oligarch-raider, accused of having conducted a massive raiding campaign over the roughly ten years up to 2010."[12]

Given all this, it should come as no surprise that Kolomoisky's business practices have long received the attention of U.S. law enforcement officials. Because of his activities, he was eventually placed on a U.S. government visa ban list, prohibiting him from entering the country legally.[13] As we will see, he would be taken off that list shortly after Hunter Biden and Devon Archer joined the board of his energy company.

It is important to understand the events in Ukraine that were unfolding as Biden and Archer, who was on the board of the Heinz Family Office, joined Burisma. It may also explain why they were invited to join.

In February 2014, a series of protests and strikes across western Ukraine culminated in a political revolution. President Yanukovych was chased out of office, implicated in both rampant corruption and shameless brutality. He was also very cozy with the Kremlin. (He was eventually exiled to Moscow.) The wave of protests brought into power a coalition government that was nationalist and determined to loosen the country's ties to Russia.[14] Vladimir Putin, sensing an opportunity, pushed Russian military forces into the Crimean Peninsula and then the easternmost region of Ukraine. He also armed pro-Russian militias in eastern Ukraine who savagely fought the new Ukrainian government.[15] Crimea is a strategic peninsula that offers a warm water port into the Black Sea and has long been a sought-after Russian prize.[16]

The Ukrainian government responded by mobilizing the army. There was vicious fighting in the eastern region of the country, including Dnipropetrovsk. Burisma has many of its energy assets in the area. Kolomoisky was appointed governor of Dnipropetrovsk. Never one to leave his sentiments unknown, Kolomoisky offered a $10,000 bounty for each Russian "saboteur" caught by the public. Putin called the billionaire a "unique imposter." Kolomoisky fired back, calling the Russian president "a schizophrenic, short in stature."[17]

The international community rightfully reacted to Putin's move with shock and anger.[18] The response from Washington was almost immediate.[19] And for Biden and Archer, there was again the opportunity to strike deals in the wake of the vice president and secretary of state's official duties.

In early March, only days after the Russian move into Crimea, Secretary of State John Kerry visited Kiev, arriving with a pledge of $1 billion in American loan guarantees and offers of technical assistance. He also announced clear-cut American political and moral support for Ukraine. As he walked along Instytutska Street in the heart of Kiev, a Ukrainian woman beseeched him, "We hope Russian troops will leave Crimea, and we also hope for your assistance."

"We are trying very hard," Kerry responded.[20]

Kerry spoke forcefully about the U.S. commitment to an independent Ukraine. But it was Vice President Biden who would end up being "point person" in the Obama administration's policy toward Ukraine.[21] "No one in the U.S. government has wielded more power over Ukraine than Vice President Joe Biden," noted *Foreign Policy* magazine.[22] Indeed, his power as it relates to Ukrainian policy extended far beyond just Washington; he was "considered the voice of the country's western backers."[23] Biden consulted regularly with the Ukrainian president by telephone and made five trips to the Ukraine between 2014 and 2017.[24] He did so at the same time that his son and his son's business partners prepared to strike a profitable deal with controversial and reportedly violent oligarchs, Kolomoisky and Zlochevsky, who would benefit from his actions.

On April 16, 2014, Devon Archer made a private visit to the White House for a meeting with Vice President Biden. We do not know the duration because, according to White House records, the meeting lasted until 11:59 p.m., the end-of-the-day placeholder when the meeting's end was not recorded.[25]

Less than a week later, on April 22, there was a public announcement that Devon Archer had been asked to join the board of Burisma. Three weeks after that, on May 13, it was announced that Hunter Biden would join, too. Neither Biden nor Archer had any background or experience in the energy sector.[26]

As was the case with their deals in China, the foreign company, Burisma here, did not hide the fact that the son of the vice president and the financial manager for the family of the American secretary of state were joining the board. Far from it. They mentioned it in the first paragraph of their press release (as translated): "Devon Archer is an iconic figure in American politics today. According to US media, he is the part of the Family Foundation Directorship Heinz Family Office with Christopher Heinz, John Kerry's stepson (the current US Secretary of State). He has served as Senior Adviser to John Kerry and in 2004 during the presidential campaign. Today he is the co-founder and managing director of Rosemont Seneca Partners, where his partner is Hunter Biden, the son of the current US Vice-President Joseph Biden."[27]

The timing of the announcement is significant. The day before Archer's appointment, on April 21, Vice President Joe Biden landed in Kiev for a series of high-level meetings with Ukrainian officials. The vice president was bringing with him highly welcomed terms of a United States Agency for International Development (USAID) program to assist the Ukrainian natural gas industry, and promises of more U.S. financial assistance and loans. Soon the United States and the International Monetary Fund would be pumping more than $1 billion into the Ukrainian economy.[28]

The younger Biden, for his part, tried to put the best possible face on the deal. He claimed that by joining the board of the scandal-laden natural gas producer, he would "contribute to the economy and benefit the people of Ukraine."[29]

Biden also said that his role would be "consulting the Company on matters of transparency, corporate governance and responsibility, international expansion and other priorities." His responsibilities also included providing "support for the company among international organizations."[30] At other times they would describe his role as offering "strategic guidance to Burisma."[31]

Biden and Archer's appointment to the Burisma board did not go unnoticed in China, where Rosemont Seneca enjoyed continuing progress with its BHR investment fund with the Chinese government. The Chinese government media ran several stories about Biden's Ukrainian venture. The main Chinese government news outlet *China Daily* noted, "The decision immediately sparked speculation over whether the US government had a role in Hunter Biden's promotion, as his father has frequently spoken about the need to increase Ukraine's energy independence."[32]

Another ally of Secretary of State Kerry was soon added to the Burisma payroll. Less than a month after Kerry's protégé Archer joined the board, Burisma hired a new lobbyist in Washington named David Leiter, a former Senate chief of staff for Kerry.[33]

Burisma's access to the corridors of power halfway around the world in Washington was looking bright.

At the end of June, four U.S. senators (Senators Markey, Shaheen, Wyden, and Murphy) drafted a letter to President Obama requesting greater U.S. aid for the Ukrainian energy industry. In a statement, Burisma immediately "applauded the range of U.S. legislative support for development of Ukraine's broad and untapped resources and an increase in transparency and good governance."[34] In 2016, Burisma organized and cosponsored Ukrainian president Petro Poroshenko's meeting with top U.S. lawmakers in Washington. In an event in the congressional auditorium at the U.S. Capitol

Visitor Center, Devon Archer joined Poroshenko and members of Congress and the U.S. Senate to discuss Ukrainian issues.[35]

The choice of Hunter Biden to handle transparency and corporate governance for Burisma is curious, because Biden had little if any experience in Ukrainian law, or professional legal counsel, period. He seemed undeterred by the fact that as he was joining the Burisma board the British government's Serious Fraud Office (SFO) was seizing $23 million from Zlochevsky's bank accounts. The SFO was also seeking access to his Swiss bank accounts and the company's account at BNP Paribas in London.[36] British officials aggressively pursued the case for more than a year but ran into a roadblock when Ukrainian prosecutors refused to cooperate in the investigation. By January 2015, the SFO had unblocked the accounts.[37]

Doing business in the Ukraine is not for the faint of heart. According to the World Economic Forum, Ukraine is one of the most corrupt places in the world. Out of 148 nations studied, it ranks

#122 for "diversion of public funds"
#143 for property rights
#130 for "irregular payments and bribes"
#133 for "favoritism in decisions of government officials" (i.e., cronyism)
#146 for "protection of minority shareholders' interests"[38]

As one Reuters writer put it, "Corruption in Ukraine is so bad, a Nigerian prince would be embarrassed." Indeed, Ukraine was rated as more corrupt than Nigeria by Transparency International.[39]

But even in such a corrupt climate, Burisma seems to stand

out. Biden and Archer were joining one of the most corrupt firms in a very corrupt country. One year after Hunter Biden joined the firm to offer help on compliance and legal matters, experienced industry observers warned investors that Burisma was still a company to be avoided. "Anyone considering investing in Ukrainian gas or this influential player [Burisma] should understand that the next stop [for their money] will be Switzerland," declared one industry publication. It went on: "Burisma—registered in Cyprus in 2006—fails to pass the most basic due diligence check. Its registration documents are impossible to run down. It publishes no asset information or financial records nor does it release any audited financial statements. The complete lack of transparency means that anyone interested—including potential investors—must rely solely on press releases about Burisma's future plans and intentions."[40] It was hardly what you would call a ringing endorsement.

Yaroslav Udovenko, managing director of the investment firm Empire State Capital Partners in Kiev, says there is a very short list of legitimate gas companies in Ukraine. "Burisma isn't on it."[41]

Despite the persistent legal questions swirling around Burisma, Archer and Hunter Biden carried the flag around the globe, serving to legitimize the activities of the company. Devon Archer represented Burisma at the Louisiana Gulf Coast Oil Exposition in late 2015. Hunter Biden attended and addressed the Energy Security for the Future conference in Monaco. Sponsored by Prince Albert II of Monaco, Biden joined him onstage along with former vice-chancellor of Germany Joschka Fischer.[42]

How much were Biden and Archer paid by Burisma? There is no way of knowing. They are not required to disclose their compensation. One can only imagine that lending one's name to a suspect company would require a large sum. Given the large sums businesses pay for advertising, public relations, and marketing, one wonders how much Burisma would invest in gaining credibility by appointing two very politically connected Americans to their board.

Professor Oliver Boyd-Barrett, who has written about Ukraine, estimated that the deal with Burisma could be enormous. "Potentially," he wrote, "the Biden family could become billionaires."[43]

While we cannot trace the sums of money going to Hunter Biden and Devon Archer, we can map the funds and benefits flowing to Burisma's owners at the direction of Hunter Biden's father, Vice President Joe Biden, and Devon Archer's confidant, Secretary of State John Kerry. The transactions raise questions not only because of their timing but also the large sums of money involved.

Recall that controversial oligarch and Burisma owner Kolomoisky had been barred entrance into the United States because of legal concerns about his activities in the Ukraine. But in 2015, after Biden and Archer joined the board, that changed. With the intervention of the U.S. embassy in Kiev, he was given admittance back into the United States.[44]

Then there was the disappearance of $1.8 billion in U.S. taxpayer-guaranteed money. Both Joe Biden and John Kerry championed $1.8 billion in taxpayer-backed loans to be given to Ukraine courtesy of the IMF.[45] The funds were being loaned to the country to keep the country's financial markets liquid. Much of that money would go through Kolomoisky's PrivatBank. And more than $1 billion from Privat just simply disapppeared.[46]

Where did that money go?

We can actually follow the egress of this large sum of money into offshore private accounts, thanks to investigative work by a Ukrainian anticorruption watchdog group called Nashi Groshi

("Our Money"). Nashi Groshi mapped the complex flow of money by researching a series of court judgments of the Economic Court of the Ukraine's Dnipropetrovsk region, where Kolomoisky was governor until he left the country. It "worked like this," the watchdog says,

> Forty-two Ukrainian firms owned by fifty-four offshore entities registered in Caribbean, American, and Cypriot jurisdictions and linked to or affiliated with the Privat Group of companies, took out loans from PrivatBank in Ukraine to the value of $1.8 billion. The firms then ordered goods from six foreign "supplier" companies, three of which were incorporated in the United Kingdom, two in the British Virgin Islands, one in the Caribbean statelet of St. Kitts & Nevis. Payment for the orders—$1.8 billion—was shortly afterwards prepaid into the vendors' accounts, which were, coincidentally, in the Cyprus branch of PrivatBank. Once the money was sent, the Ukrainian importing companies arranged with PrivatBank Ukraine that their loans be guaranteed by the goods on order.[47]

Of course, it was all a scam, and the complex corporate structure a decoy screen. No "supplier" companies provided any goods. The $1.8 billion effectively disappeared from Privat, laundered through a network of offshore entities. "This transaction of $1.8 billion . . . with the help of fake contracts was simply an asset siphoning operation," the watchdog explained.[48]

By December 2016, Ukrainian authorities were forced to nationalize Kolomoisky's PrivatBank when it was discovered that the bank was collecting savings from Ukrainians and then engaging in "massive insider lending" whereby mysterious front corporations were taking large loans and then not paying them back.[49]

Huge sums of money had simply disappeared. The Ukrainian government, supported by taxpayer dollars from the West, was forced to take ownership of the bank and make sure customers did not lose their money. One Ukrainian lawmaker called it the "greatest robbery of Ukraine's state budget of the millennium."[50]

When money disappears, the people who lent that money want answers. And so at the behest of angry officials in the United States and the West, an investigation was launched.

Burisma company founder Mykola Zlochevsky, the stocky former environmental minister, worked deftly to avoid criminal charges. In February 2016, Ukrainian authorities seized his property on suspicion that he had engaged in "illicit enrichment."[51] Zlochevsky fled the country and was placed on Ukraine's wanted list.[52] The Ukrainian Prosecutor General's Office actually seized Burisma's gas wells. Tax authorities began investigating him for suspicion of tax evasion.[53]

But then, almost as quickly as the investigation into Zlochevsky and Burisma was launched, it was halted.

On January 16, 2017, Air Force Two was descending on Boryspol Airport, just southeast of Ukraine's capital.[54] This would be Joe Biden's last foreign trip before leaving office. It was cold and dark. Dressed in an overcoat, he descended the stairs quickly and was met by a delegation from the Foreign Ministry on the tarmac. This was his Ukrainian "swan song," as Reuters put it, "a farewell visit by one of Ukraine's strongest political supporters." The Obama administration, under Biden's direction, had poured some $3 billion into the country.[55] Later that evening he met with the prime minister of Ukraine, and then a late-evening meeting with his friend Petro Poroshenko, the Ukrainian president. As the *Kyiv Post* reported, that latter meeting was to be "behind closed doors, and details of most of their discussion won't be made public."[56]

Four days before Biden arrived, Burisma made a dramatic announcement: the Ukrainian criminal investigations into the company and its founder had been ended by Ukrainian government prosecutors. "Since all legal proceedings against Burisma Group are closed," the company announced, "it will allow us to increase production volumes and the flow of foreign investments in Ukraine, consider attracting international companies in the country, fulfill social and investment responsibilities, as well as duly pay in full all required tax liabilities in the budget. This is a big step forward for Ukraine in general and Burisma Group, in particular."[57] Joe Biden met with Ukrainian president Poroshenko and was warmly greeted by Ukrainian officials who saw him as their strongest advocate in the West. Biden urged the United States and the West to stand behind the Ukrainian government: "The international community must continue to stand as one against Russian coercion and aggression," he told reporters, standing beside Poroshenko.[58]

Despite the enormous evidence of corruption and criminal activity involving Burisma, Devon Archer remained involved with the company through the end of 2016, while Hunter Biden as of this writing still serves a member of the board and provides legal assistance. They have never disclosed their compensation. And the more than $1 billion that disappeared from their business partner's bank has never been recovered.

The Pine Ridge Indian Reservation in South Dakota is a world away from the Ukraine, but the story here helps put Archer's global activity, and Rosemont Seneca's methods, in context. Home to one of the poorest Native American tribes in the United States, the Oglala Sioux, it has struggled over the years to attract investment and business opportunities for tribal members.[59] Beginning in 2014, the tribe was fleeced courtesy of a complex, fraudulent scheme involving a known con artist, a man once proclaimed "Porn's New King," and Kerry confidant and Rosemont Seneca partner Devon Archer. It was an audacious scheme. For more than two years, Archer worked with a handful of other individuals to extract more than $60 million in bond money from the tribe. Proceeds that were supposed to go to the tribe ended up being used to purchase luxury goods and line the coffers of the scammers.[60]

Caught up in the scheme was a company called Burnham Asset Management. Hunter Biden served as the vice-chairman of the company. Devon Archer sat on the board of directors. Millions would flow through an entity run by Archer called Rosemont Seneca Bohai.[61] It would end with the May 2016 arrest of Devon Archer by federal agents.[62]

In March 2014, a large man with a long record of fraudulent financial schemes ventured to Las Vegas, Nevada, to attend the National RES (Reservation Economic Summit).[63] It was an economic development conference for Indian reservations from across the country. John Galanis, with a heft of nearly four hundred pounds, had been behind the pump and dump stock deal involving the "Hair Extension Center" and a real estate deal in which he bilked Eddie Murphy and Sammy Davis Jr. out of money. In 1988 he was sentenced to twenty-seven years in prison for tax shelter fraud. In 2000 *Forbes* magazine called him "one of the biggest white-collar criminals of recent decades."[64]

His son Jason was reportedly involved in investment schemes of his own. In 2003, he was dubbed "porn's new king" for his early online-porn payment-processing site.[65] According to *Forbes* magazine, while in jail John was directing Jason's financial activities and was calling his son collect thirty times a day to discuss business matters.[66]

In Vegas, the Galanises sat down with the leaders of the Wakpamni Lake Community Corporation, a tribally chartered economic development corporation owned by the Oglala Sioux. They suggested that the tribe issue bonds to raise capital for economic projects. Tribal leaders were eager to invest in a winery, among other ventures.[67]

With a deal in the making, Jason reached out to Devon Archer and another finance guy, Bevan Cooney, to offer them a part of it. On April 4, 2014, Jason wrote them an e-mail with the subject line: "Oglala Native Spirits Memo.docx." The body of the e-mail said, "$20m bond approved. Proceeds are $15mm to us and 5mm to them for a winery investment they want to make."[68]

Devon Archer, and his association with Biden and Kerry, would prove to be an attractive, legitimizing factor for the Galanises' activities. As Jason Galanis sought investors, he would e-mail promotional materials to use "when appropriate to demonstrate who your financial sponsors are." Under "Leadership," the first name to appear was Devon Archer, "Managing partner of Rosemont Group, a $2.4 billion private equity firm co-owned by Hunter Biden and Chris Heinz." It went on to note that he was "Vice-Chairman and Investment Committee member of Bohai Harvest RST."[69]

In August, the first bond issuance worth $27 million was released.[70] Less than two months later, on October 1, a second bond issuance of $15 million took place. Who bought those bonds? According to the SEC, the proceeds were diverted from the first bond sale to buy into the second, using Rosemont Seneca Bohai, LLC, to do it.[71]

Around August 15, 2014, Jason Galanis, whose nickname was "Greek," sent another e-mail to Archer and Cooney about the legal execution of the bond sale. "This is pure genius alla [*sic*] mikey Milken!!" wrote back Cooney. "The Native American Bonds! . . . Great work here Greek!"

On September 24, 2014, $15 million was transferred into the Rosemont Seneca Bohai brokerage account in New York from an entity called Thorsdale. But the money did not stay there long. On October 1, Rosemont Seneca Bohai, LLC, purchased the entirety of the second tribal bond issuance for $15 million.[72] When a brokerage firm associated with the deal asked where the $15 million had come from, the government alleges that Archer lied and said, "The funds used to purchase the bonds were from real estate sales through my business Rosemont Seneca Bohai, LLC." But of course, that was not true. The funds had been misappropriated from the first tribal bond issuance.[73]

A little more than six months later, on April 9, 2015, Rosemont Seneca Bohai, LLC, transferred those $15 million in tribal bonds into an obscure brokerage account in Bermuda called VL Assurance. As U.S. prosecutors would point out, "Rosemont did not receive anything of economic substance, such as an equity interest in VL Assurance (Bermuda) Ltd., in exchange for this transfer."[74]

According to federal investigators and prosecutors, the proceeds of these bond sales were misappropriated and went into various projects. Money that was supposed to go to the tribe ended up being used for the benefit of the Galanises and Devon Archer. Some of the proceeds went into a legal defense fund for John and Jason Galanis, who had legal bills related to another financial scheme. Some went to luxury items. The Galanises poured tribal bond money into purchasing high-priced items from Prada, Gucci, and Valentino, among others. Other money ended up in Devon Archer's pockets, including over $700,000.[75] Some ended up in the accounts of Rosemont Seneca Bohai.

Some of that money may have found its way to another Rosemont entity.[76]

Rosemont Seneca Technology Partners, which listed Hunter Biden as an adviser and Chris Heinz as an investor, created a joint venture with the state of Hawaii called mbloom to incubate high-tech companies in the islands.[77] On May 19, 2015, a small high-tech company called Code Rebel had an IPO on the Nasdaq Stock Market. The company was headed by Arben Kryeziu, a Kosovo-born German citizen who was running mbloom.[78] As alleged by the SEC, Jason Galanis and the two others used tribal bond proceeds to buy up to 87 percent of the stock on that day, driving the stock up from its $5 offering price to $15 a share. Buying the remaining shares were Devon Archer and a few other investors.[79]

The scheme began to unravel when tribal leaders wondered where the money was going. Federal authorities launched an investigation and on May 11, 2016, Devon Archer and the others involved in the scheme were arrested and charged with "conspiracy to commit securities fraud."[80]

They had also used pension funds to buy the bonds and quickly faced civil cases as well. The pension fund for workers at Michelin, the tire company, sued Archer and the others in court alleging that the defendants had "raided the proceeds from closing the bonds in order to create a 'slush fund' that would pay certain Defendants, directly and indirectly, to support their lifestyles, legal fees, and speculative ventures." According to the criminal complaint, $700,513 ended up going to Archer via the Rosemont Seneca Bohai, LLC, account.[81]

Both the criminal charges and the civil lawsuits name Devon Archer, but no one else connected with Rosemont Seneca. Did Archer go rogue? While there is no direct evidence that Hunter

Biden or Chris Heinz knew of the tribal bonds scheme, the corporate structure of their business together suggests that they could have—and maybe should have—known.

In the criminal complaint, prosecutors noted that Archer was the founder of Rosemont Capital, LLC, and was connected with Burnham Asset Management. Hunter Biden's name was not mentioned. The business that was used to transfer the money was Rosemont Seneca Bohai. According to legal records in New York, Rosemont Seneca Bohai uses the same business address as the other Rosemont entities involving both Biden and Heinz. It is also interesting to note that Rosemont Seneca Bohai, LLC, is also listed as a shareholder on the Bohai Harvest RST deal in China.[82] There were large sums of money—sometimes $15 million—passing through the Rosemont Seneca Bohai bank accounts.

Hunter Biden was a partner in Rosemont Seneca. He was also the vice-chairman of Burnham. Did he not notice?

Devon Archer was arrested in May 2016 for his alleged role in the tribal bonds scam. The Ukrainian News Agency story said it all: "Burisma Holdings Board Director Archer Arrested," blared the headline.[83] As of this writing, he is awaiting trial. (Jason Galanis has been sentenced to fourteen years for his involvement in the scam.)[84] Following Archer's arrest, his name was removed from the Burisma board. Hunter Biden remains on the board and as an adviser to Burisma.

The environs of the White House are not the only place you will find American Princelings. They also roam the highest levels of Capitol Hill.

5

McCONNELL AND CHAO: FROM CHINA WITH PROFITS

- Senator Mitch McConnell and his wife, Elaine Chao, benefit from close ties to the Chinese military-industrial complex.
- Chao's family has reaped large profits thanks to the Chinese government.
- McConnell has become less critical of Beijing as those relationships have blossomed.

Senate majority leader Mitch McConnell and his wife, Elaine Chao, operate at the highest levels of government. They are one of the most powerful couples in the United States, with McConnell climbing to become the senior senator in the United States, while his wife, Chao, has been a member of the cabinets of two presidents: secretary of labor under George W. Bush and secretary of transportation under Donald Trump. She previously was the head of the Federal Maritime Commission. Along the way, she has served on numerous corporate boards.[1]

McConnell and Chao have seen their wealth increase dramatically in a matter of just a few years. In 2004, they had an average net worth of $3.1 million according to public disclosures—well below the senate average of $14.5 million. Ten years later, they had a net worth of between $9.2 million and $36.5 million. The key: in 2008 they received a gift from Elaine Chao's father, James. The Chao family fortune comes from the Foremost Maritime Corporation (later renamed Foremost Group), a shipping firm that her father founded, and which remains a family business. James Chao is the chairman, and Elaine's youngest sister, Angela, is deputy chairwoman, running the day-to-day operations. Another sister, Christine, is Foremost's general counsel. Elaine worked for the company in the 1970s. "Shipping is our family tradition," Elaine said in a 2016 speech at the National Taiwan Ocean University.[2]

So what family role do McConnell and Chao play, exactly? As these two have held court in Washington, Foremost has thrived, thanks in large part to close relations with the Chinese government. As McConnell and Chao have risen in political stature, these relationships have only strengthened.

Senator McConnell's father-in-law and sister-in-law have served on the board of directors of a Chinese government-owned military contractor. Another sister-in-law has been appointed to the government-controlled Bank of China, only the second foreign national to serve on that board. The Chaos' shipping company has done large volumes of business with the China State Shipbuilding Corporation (CSSC). While Foremost is based out of the United States, their ships have been constructed by Chinese government shipyards, and some of their construction financed by the Chinese government. Their crews are largely Chinese. (Ironically, Elaine Chao as the U.S. secretary of transportation has maintained that ships crewed by Americans are "a vital part of our national security.") And their cargo vessels have done considerable business with Chinese state-owned companies.

The story of the Chaos has all the elements of a classic immigrant success story. James Chao came to the United States to continue his education in 1958.[3] His wife, Ruth Mulan Chu, and three daughters, including eight-year-old Elaine, joined him in 1961.[4] In 1964, James Chao launched Foremost Maritime Corporation.[5]

Once in the United States, the Chaos welcomed three more daughters.[6] They all excelled academically, each attending elite colleges. As the family grew, so did the business.

Elaine became a White House fellow during the Reagan administration and rose to become the chairman of the Federal Maritime Commission.[7] In 1987, she met a relatively new senator named Mitch McConnell who had taken office in 1985 to represent the state of Kentucky.[8] The Chao family began donating heavily to McConnell's campaigns. Mitch and Elaine started dating in 1991 and were married in 1993. Elaine describes the relationship as a bit of a surprise. "I thought I would marry a nice Chinese boy," she recounted for one reporter, "but I'm too tall."[9] So she found a nice senator from Kentucky. Still, James Chao remained the traditional head of the family. "When there is a family dinner, no one touches the food before James Chao," not even Senator McConnell.[10]

The wedding was a traditional affair. Senator McConnell described it as "an exceedingly joyful day." But *who* attended spoke of what would lie ahead. Wedding attendees included family, friends, and the Chinese representative to the United Nations, a family friend.[11]

Months after the wedding, in December 1993, Senator McConnell found himself in Beijing for a series of private meetings with the most senior officials in the Chinese government. This was not a congressional trip or something arranged by his senate office. The meetings were arranged by McConnell's new father-in-law, James Chao, and came at the invitation of the China State

Shipbuilding Corporation (CSSC). McConnell met with Chinese president Jiang Zemin, a classmate of James Chao at Jiao Tung University in China decades ago.[12] There was also a private meeting with the Chinese vice-premier, Li Lanqing.[13]

So why was McConnell in Beijing for these meetings? Were the meetings political or commercial? The answer is they likely were both. As is so often the case in Beijing and Washington, politics and business are deeply intertwined.

At the time, in 1993, China was isolated internationally. In June 1989 the Chinese government had sent thousands of troops into Tiananmen Square, leading to the massacre of an unknown number of Chinese students—perhaps over five hundred.[14] An estimated ten thousand people were arrested.[15] McConnell was only the second Republican U.S. senator to visit the country since the massacre; the Chinese government made sure to publicize his visit.[16]

But this also had the appearance of a business meeting. According to the Chinese government, McConnell and the Chao family "arrived [in Beijing] at the invitation of the China State Shipbuilding Corporation."[17] In 1997, there was another private meeting among Jiang, Senator McConnell, and Elaine Chao, this time around the official state Chinese dinner in Washington. There would be future private meetings with the Chinese leaders who would follow Jiang.[18]

The CSSC would come to play a central role in the family's financial success, and Mitch McConnell would increasingly avoid public criticism of China.

As the Chaos and the Chinese government went into business together, the Chaos-McConnells tied their economic fate to the good fortunes of Beijing. Were McConnell to critique Beijing aggressively or support policies damaging to Chinese interests, Beijing could severely damage the family's economic fortunes.

The evolution of Senator McConnell's views on China are on public record. In 1989, just after the Tiananmen Square massacre, McConnell gave a hard-line speech at the University of Louisville. McConnell was a member of the Senate Committee on Foreign Relations at the time. "We'll never forget the sight of those young people without arms up against tanks and machine guns," he said. Later that year it was revealed that then president George H. W. Bush covertly sent two close aides, including National Security Advisor Brent Scowcroft, to meet with Chinese officials. McConnell was the only Republican in the Senate to criticize the mission.[19]

In 1991, with free and democratic Hong Kong set to be turned over to Chinese control in 1997, McConnell sponsored the U.S.–Hong Kong Policy Act of 1992. The law "called on China to preserve Hong Kong's market economy once it returned to Beijing's control in 1997 and preserved a separate immigration quota for Hong Kong after 1997."[20]

McConnell's rhetoric and policy toward mainland China softened in the decades that followed. When Hong Kong's turnover approached in 1997, he reassured people like conservative writer Cal Thomas that China would become more free. Thomas explained that as McConnell saw it, "in future years, it is probable that the People's Republic of China will look more and more like Hong Kong and not the reverse." McConnell went on talk shows like *Hardball with Chris Matthews* on MSNBC and declared that the United States needed to be "ambiguous" as to whether we would come to the defense of Taiwan if attacked by China. "I think we're purposefully ambiguous. I think that's the place we ought to be."[21]

In 1999, Senate Committee on Foreign Relations chairman Senator Jesse Helms introduced the Taiwan Security Enhancement Act. It was a show of support for a Taiwan independent of Chinese control. The bill had twenty-one cosponsors and heavy Republican support. But McConnell was not on the list.[22]

Central to China's trade strategy was to get rid of a requirement that Congress had put in place following the Tiananmen Square massacre for annual certification that the country was making progress on human rights to retain its trade status. In 2000, McConnell cosponsored S.2277, which would do just that.[23] The Chinese government had been fighting to get the requirement removed for years.

More recently, Senator McConnell has worked to upend legislation that would be damaging to Beijing. In September 2011, the Currency Exchange Rate Oversight Reform Act (S.1619) was introduced in the Senate. The bill punished countries with "fundamentally misaligned currenc[ies]" by tacking on import duties.[24] While it did not explicitly target China, the bill was considered a direct response to China's undervalued currency, which threatened U.S. competitiveness in the international marketplace and made U.S. goods more expensive in China.[25] McConnell was adamantly opposed to the bill. On October 6, 2011, he attempted to waylay the senate majority leader, Senator Harry Reid, from taking the bill to a vote, by suspending rules on the Senate floor and introducing amendments. The two rivals got into a heated, "unscripted" debate when McConnell supported seven amendments to kill the bill.[26] Reid then used a version of a rare procedure called the "nuclear option" to change the Senate rules that passed by the Democratic majority of the Senate and left McConnell "fuming."[27] On October 11, 2011, the bill passed with a 63–35 vote. McConnell voted against the measure.[28]

McConnell has defended his record regarding China by saying that he has been a consistent free trader regardless of the country.

"I was a free-trader long before I met Elaine," he told one newspaper, "and I think I've been on the free-trade side of virtually every issue, not just related to China."[29]

But his record does not reflect that. On the issue of currency manipulation, he voted to impose sanctions on Japan for manipulating its currency and allegedly engaging in unfair practices to help their automobile industry. He has supported trade restrictions on Burma on account of its human rights record and has called for protections of certain industries like tobacco during the Trans-Pacific Partnership. There is nothing in his previous record concerning any deals involving China.[30]

Likewise, Elaine Chao has been accused over the years of taking a soft position as it relates to China, and has avoided criticizing the government. During the 2008 Beijing Olympics, eight Americans were arrested for trying to protest the Chinese occupation of Tibet. Chao was serving as the official U.S. representative to the Olympics. She publicly lauded the Olympics as "a unique opportunity for the Chinese people to demonstrate the progress they have made and their sincere desire to engage with the world at every level," but never mentioned the arrests of the Americans or numerous human rights problems in the country. Anne Applebaum of the *Washington Post* criticized her comments as something that would "reinforce the Chinese regime's legitimacy among its own people, cover up its bad record and buff its image around the world—which was precisely what the Chinese regime had hoped people like her would do all along."[31]

While Chao was the Secretary of Labor in the George W. Bush administration, her department took positions that were beneficial to the Chinese government as well. The Department of Labor resisted efforts to call out the Chinese government over its workers' rights practices. When a petition was filed with the federal government under Section 301 of the U.S. Trade Act of 1974 regarding workers' rights in the People's Republic of China, she opposed it. Instead, she pointed to the fact that her department had given millions of dollars in taxpayer grants to the Chinese government to provide workplace education programs.[32]

One could argue that in each of these instances Chao was a representative of the U.S. government, serving in a cabinet; she was echoing the policy line of the administration, not her own views. But you cannot make that argument when she traveled to China as a private citizen. In 2013, she accompanied her father to Peking University where he delivered a speech entitled "American in Spirit, Chinese at Heart." Elaine Chao answered questions from students after his presentation. She never mentioned any issues related to human rights or freedom of speech. "China has to find its own path," she told students, according to the university's version of the exchange. "You cannot adopt the American solutions without adjustment. China has to find its own solution to its own problems, in its own way. You are going to be the leaders of China for the future. And so you should take pride in your Chinese culture, in your Chinese heritage, and help your country find its own path forward."[33]

She gave a curious interview to a Chinese-language newspaper called *Sing Tao* in July 2000 under her Chinese name, Chao Hsiao-lan. In an article titled "Washington Penetrating Inspection—Exclusive Interview with Chao Hsiao-lan," Chao spoke openly about a number of important issues between the United States and China. At the time of the interview, there was a lot of discussion in Washington about the Cox Report, a bipartisan congressional report on Chinese espionage against the United States and a series of illegal financial contributions made by Chinese officials to politicians in the United States. Chao was critical of the report and concerns about Chinese meddling in American politics. She made clear she "in no way" agreed with the report's findings. And she went on to dismiss the idea that China could pose any threat to the United States. "The U.S. is always happy to see an enemy, possibly because it serves a domestic policy purpose. And now that the Soviet Union doesn't exist and Russia is no longer the evil empire, some people without any reason make China the United States' enemy." She went on to criticize Democrats who raised questions about the human rights situation in China and who were worried about trade imbalances. She added, "Unfortunately, the Republican party also has criticizers of China who are Christians and religious believers. They also talk about human rights and think China is an enemy, but compared to the Democrat side, they are apt to be well organized." She urged the Chinese government to fight a public relations battle in the United States. "China also must come up with their own persuasive formula to win the public-relations war. I'm saying to China, this is a new game China may have difficulty understanding. I believe China can do well."[34]

As McConnell gained political weight on Capitol Hill and his position toward China softened, the Chao family saw its business ties with the Chinese government increase dramatically—to their financial benefit. Recall that as we saw earlier, China uses commercial ties to curry favor with foreign politicians. As scholar Jeffrey Reeves notes, China works to develop ties with a country's "political elite" and hopes to leverage "elite relations" to advance their interests in those countries.[35] In Washington, it does not get much more elite than McConnell and Chao.

Very few in Washington political circles know the extent to which the McConnell-Chao family is financially linked with Beijing. Indeed, the extent of their ties never comes up. But some of those ties are well-known within the small circle of global shipping companies. As the shipping industry publication *TradeWinds* puts it, "Industry players describe Foremost Maritime as a low-profile shipping company with strong ties to China. They also say it has links with both the US and Chinese governments."[36] It is a phrase you see repeated over and over again in trade publications about Foremost: "close links with both the US and Chinese governments."[37]

The links with the U.S. government, of course, are clear to us Westerners. Both Senator McConnell and Elaine Chao hold senior positions in government. Troubling links with the Chinese government are far murkier if not obscured completely from the western view.

As McConnell and Chao were beginning their marriage, two trajectories developed simultaneously. Senator McConnell grew in seniority in the Senate, and the Chao family became increasingly intertwined financially with the Chinese government. The China State Shipbuilding Corporation describes James Chao as a "business partner and friend of years as well."[38] Foremost was "a good client of China State Shipbuilding Corp," according to industry insiders.[39]

Before McConnell and Chao married, the family had two large cargo ships built by the Jiangnan shipyards in China, the *Hsing May* in 1990 and the *Yu May* in 1991.[40] But ten years later, by the summer of 2002, the Chao family was doing increasing volumes of business with the Chinese government and the government's Shanghai Waigaoqiao Shipbuilding Company, which produced for them Capesize ships, the largest dry cargo vessels in the world. (These ships are so big, they cannot transit the Suez Canal.) Foremost Maritime would have CSSC construct six of the massive vessels at the yard over the next four years.[41] Between

2001 and 2011, the company delivered ten of the mammoth ships to Foremost.[42]

Teh May was built by Shanghai Waigaoqiao in 2004.[43] They added the *An May* in 2005, built by the same shipyard.[44] By 2006 they had ordered their sixth massive ship from the Shanghai Waigaoqiao Shipbuilding yard. They had the *Guo May* built in 2011.[45] By 2014 the Chaos had ordered six 180,000-dwt (deadweight ton) bulkers at another government-owned shipyard, Qingdao Beihai Shipbuilding Heavy Industry Company. They cost $55 million apiece.[46] Beihai reportedly sweetened the deal with the Chaos when it "squeezed out an early berth slot in late 2015 to accommodate one for the latest Foremost" ships.[47]

Having your ships built in China in itself is no big deal. But the Chaos became increasingly dependent on Beijing for their business success. The crews on their vessels are largely Chinese.[48] And they operate out of Chinese ports carrying large amounts of raw materials in and out of China. Their clients include well-known Western companies like Cargill, but often Foremost is carrying goods for Chinese government-owned entities like Rizhao Steel and Wuhan Iron & Steel (Wisco).[49]

But the ties became even closer.

Soon, both James Chao and sister Angela were appointed to the board of directors of the Chinese State Shipbuilding Corporation Holdings, Ltd.[50] You would not know it at first glance by looking at the company's documents because they used modified names in corporate records. James Chao is listed as Zhao Xicheng (赵锡成). Angela Chao is listed as Zhao Anji (赵安吉).[51] They joined the board just as the U.S. Senate was taking up sensitive legislation concerning China.

CSSC Holdings is an offshoot of CSSC. It was formed to get "private capital to enter China's defense industries." According to the annual reports of CSSC Holdings, the company is subsidized by the government. And the company's business includes doing military work. At the time that they joined, the Chao family members appear to have been the only foreigners to serve on the company's board.[52]

As a state-owned defense conglomerate, CSSC is at the heart of the Chinese government's military-industrial complex. CSSC and China Shipbuilding Industry Corporation (CSIC) are together "the world's most prolific builders of large surface combatants and submarines."[53] The CSSC is not shy about its role, describing itself as an "extralarge [*sic*] conglomerate and state-authorized investment institution directly administered by the central government of China."[54] The former chairman of the company, Hu Wenming, declared that the company's "number one priority" is building "military products" and he vows the company will continue in "strengthening the [Chinese] military." He adds that "CSSC's cadres and workers are deeply encouraged and inspired to take on the responsibility of using 'ships to serve the country.'"[55]

CSSC, while the Chaos served on the board of directors of CSSC Holdings, played a central role in this massive Chinese naval buildup. By 2020, many naval experts believe that "China is on course to deploy greater quantities of missiles with greater ranges than those systems potentially employed by the U.S. Navy against them." And according to the China Maritime Studies Institute at the U.S. Naval War College, the Chinese shipbuilding industry "appears to be on a trajectory to build a combat fleet that could be, *in hardware terms*, quantitatively and qualitatively on par with that of the U.S. Navy by 2030."[56]

The fact that both the father-in-law and sister-in-law of Senator McConnell sat on the board of CSSC Holdings is highly unusual, to say the least. One could say it is unprecedented in

American political history. In general, CSSC is a sensitive, critical asset for the Chinese government and operates under a veil of privacy and secrecy. Details about CSSC "are very opaque," says Julian Snelder, a partner in a global investments fund at a conference organized by the China Maritime Studies Institute at the U.S. Naval War College. The company consists of a variety of "mixed civil-military . . . complexes," including sensitive research institutes that are secretly operated and report directly to the government's State Council.[57]

Senator McConnell and Elaine Chao do not have a direct ownership stake in Foremost Group. But clearly, they are beneficiaries of its financial success. In 2008, they received a "gift" from James Chao of between $5 million and $25 million. This dramatically increased McConnell's personal wealth, more than doubling it.[58] They reasonably stand to receive other "gifts" as family fortunes allow. Financial success for Foremost means financial success for McConnell and his wife.

But the money flowing from the Chinese government to the McConnell-Chao family goes beyond shipbuilding. Elaine Chao, when not serving in government, has given paid speeches to Chinese-government entities for years. It is unclear how much she has been paid for these speeches; she does not disclose her speaking fees on Senator McConnell's annual disclosure form, but her fee is as much as $50,000.[59] Speakers commonly charge more if they are required to travel overseas to deliver the address. It is impossible to know precisely how many Chinese government entities have put how much money in Chao's pockets over the years.

Repeated questions sent to Chao about these speeches went unanswered.

In 2013, Chao participated in the Chinese government's Boao Forum for Asia, which was headlined by Chinese president Xi

Jinping.[60] Event attendees included a "long list of ministers, provincial governors and party secretaries." Called the "Davos of Asia," the event is constructed so that "politicians and business leaders" who "want insights into trade with China" can meet with government officials.[61] She has also given talks to other government forums like the Forum on ICT and Urban Development in Ningbo.[62] Along the way, she collected fees for addressing numerous Chinese government-controlled universities like Tsinghua University in Beijing, Peking University, and Fudan University.[63] In 2009 she addressed the China Center for International Economic Exchanges, which is known for its links to the Chinese government.

As was the case with Bill Clinton and his speaking fees while Hillary Clinton was secretary of state, it is fair to ask how those fees might be paying for more than podium time. Paying a spouse a hefty speaking fee can be a lucrative tool for currying favor with a power broker. As Elaine Chao was being paid these fees by Chinese government entities, her husband was the Republican leader or Senate majority leader in the U.S. Senate, grappling with an abundance of legislation that would affect China's interests.

Elaine Chao's financial ties to mainland China also include corporate relationships. In the past she has sat on the board of a broadband company called MultaCom, which was offering direct Internet links between China and the United States through a joint venture with the Chinese government's China Unicom. She failed to disclose that position when she joined the George W. Bush administration as secretary of labor. The lack of disclosure was an "inadvertent omission," said Chao's spokesperson.[64]

The author repeatedly contacted the offices of both Senator McConnell and Secretary Chao about the family's close financial ties with the Chinese government. They have not responded.

During the 2016 presidential election, the Chinese government watched warily as Donald Trump charged them with engaging in a variety of unfair economic practices. Trump accused Beijing of manipulating its currency, protectionism, and stealing American intellectual property.[65] He also intimated that he saw the country as a rising military threat.[66]

When Trump won, there was disquiet in Beijing about what might lie ahead.

Ten days after Trump's win, the government-run Bank of China made an announcement. As we have seen in previous chapters, the Bank of China is deeply embedded in China's political structure. It is run by a trusted group of managers who are politically savvy. As the Congressional Research Service notes, "All of China's banks share a common governance system, involving senior bank officers, board of directors, and a board of supervisors. The senior bank officers are members of the Chinese Communist Party (CCP) and are appointed by the CCP. The officers are also assigned ranks in the Chinese government's hierarchy, ranging from the equivalent of a bureau chief to a vice minister."[67] As a result, the bank's board of directors includes all senior Chinese officials and one lone Dutch banker. But now they would be getting a new member: Senator McConnell's sister-in-law, Angela Chao.[68]

One has to ask: With the Chao family so financially linked to the Chinese government, how can this not affect McConnell's view on policy toward China? Being in favor with Beijing has built his family's empire, and losing favor could ruin the family's shipping fortunes overnight. It seems that in both Beijing and Washington, it pays to have friends (and family) in high places.

But the flourishing of Princelings on Capitol Hill goes far beyond the case of Senator McConnell.

6

THE PRINCELINGS OF K STREET

- Foreign entities and American corporations have discovered that hiring the children of powerful politicians as lobbyists is the best way to get access to them.
- Some politicians are building family empires this way, with as many as three family members cashing in.

Missoula, Montana, is a long way from Ulaanbaatar, Mongolia. The span across the globe is 5,639 miles to be exact. But if you want to get from the one to the other, the shortest route might not be to head west over the Pacific Ocean. Instead, you might try going through Washington, D.C.

Congressman Denny Rehberg was a plain-speaking rancher when he was elected to the House of Representatives in 2000. The former lieutenant governor was unopposed in the Republican primary and won the general election comfortably.[1] As a rising GOP star, he was eventually given a seat on the powerful House Committee on Appropriations, a highly prized appointment, because that body largely controls the purse strings for the federal government and determines how hundreds of billions of dollars will be spent every year in our nation's capital.

Rehberg checked all of the conventional boxes. He joined the Republican Study Committee and the Congressional Rural Caucus. He became a prodigious fund-raiser. He sat on the powerful Subcommittee on Energy and Water Development, and chaired another on health and labor issues.

But Rehberg also did something unconventional. He helped launch a small congressional group called the U.S.-Mongolia Friendship Caucus. A caucus is a voluntary group that allows congressmen with a particular interest to meet together and work on common issues related to it. The Rural Caucus, for example, includes congressmen from rural districts.

The five-member U.S.-Mongolia Friendship Caucus was established to strengthen relations between the United States and the landlocked country known for Genghis Khan between Russia and China. Why would a Montana politician launch such a caucus? Factoring in the business activities of A. J. Rehberg, the congressman's son, helps explain the seemingly serendipitous goodwill.

In 2007, shortly after graduating from Seattle University, A. J. Rehberg went to work for a Washington, D.C., lobbying firm called Gage. The firm was headed by Leo Giacometto, a former chief of staff to Montana senator Conrad Burns.[2] By 2011 the firm's lobbying clients included the government of Mongolia. According to federal records, Gage was even lobbying Congressman Rehberg's Committee on Appropriations on behalf of the Mongolian government. Their paid work included "formation of the US Mongolia Congressional Caucus" (which Rehberg of course helped launch) and "to promote trade."[3]

The required lobbying disclosure form lists Giacometto and Ryan Thomas of Gage as the lobbyists for the government of Mongolia.[4] But according to the Mongolian embassy in Washington, A. J. Rehberg was actually "the primary point of contact for the firm's work with the embassy."[5] Gage confirmed this as well in a press release.[6]

Around the same time, the congressman's son, now about four years out of college, became a vice president of a uranium company called Mongolia Forward.[7] Based out of Washington, D.C., the company was involved in a joint venture agreement with the Mongolian government's state-owned uranium company that authorized Mongolia Forward to "explore, mine and process uranium in Mongolia."[8] The goal was to import the precious metal into the United States. A. J. Rehberg had no background in the nuclear industry. His job at Mongolia Forward was to serve as vice president for "public and government affairs." Gage actually lobbied Congressman Rehberg's Committee on Appropriations on behalf of A.J.'s company, according to federal records on issues related to "exploration, mining and international trade."[9]

How much A.J. was paid by the Mongolian uranium company has never been disclosed.

A.J. also became involved in several other Mongolian business ventures, including a fiber optic network company and the first Pepsi bottling plant in the country.[10]

Doing business in Mongolia is not for the faint at heart and requires powerful connections with the government in order to succeed. It is considered one of the more corrupt countries in the world, according to Transparency International around this time. It ranks worse than China, Jamaica, and numerous African countries.[11]

As a member of the Committee on Appropriations, Congressman Rehberg was ideally positioned to help the Mongolians. In 2007 the Mongolian government signed a compact worth $285 million in U.S. taxpayer money for infrastructure projects in the country.[12] That included plans to improve a railroad that was partly owned by the Mongolian government. The purpose of the railroad would be to transport minerals—including uranium.[13]

Beginning in 2009, Rehberg also sat on the House Appropriations Subcommittee on Energy and Water Development, which meant that he had oversight of the Department of Energy (DOE) and the Nuclear Regulatory Commission (NRC).[14] The Mongolian government's hopes of importing uranium into the United States through A.J.'s company were contingent on approval from both government bodies.

In 2012, Congressman Rehberg decided to run for the U.S. Senate in Montana. He lost. Shortly after he gave up his congressional seat, he joined the lobbying firm Mercury.[15]

We have seen how the family members of vice presidents, secretaries of states, and senate leaders have benefited from foreign entities endowing them with favorable business deals in the interest of currying favor. The Rehberg case is simply a snapshot of evidence showing how this happens in Congress, too, even with relatively obscure congressmen. If proximity is power, it is hard to get closer to a politician than by going into business with a member of their family or paying a member of their family for a service. In countries like China and Mongolia, Princeling-style corruption commonly informs political decisions. It has become increasingly that way in the United States as well. As Craig Holman, an ethics expert at the watchdog group Public Citizen, puts it, "Special interests that have pending business before a member of Congress often look to throw money at the feet of a family member. That way these lobbies can still use their connections—at arm's length—to get access."[16]

Many D.C. lobbying firms spend lavishly on politicians' family members. Back in 2012, the *Washington Post* reported that "more than 500 firms have spent more than $400 million on lobbying teams that include the relatives of members [of the House and Senate]."[17] This chapter tells a fraction of the stories that those numbers represent.

Many elected officials see public office as a business enterprise, with family members orbiting around them, positioned to strike lucrative deals with foreign and American entities eager to curry favor from the powerful. In this charade of public service, bloodlines grow wealthy at the expense of public policy and the American public takes a backseat to the highest bidder.

One example very well could be Congressman Bill Shuster of Pennsylvania, who comes from a powerful political family that has been deeply embedded in politics since the 1890s. Bill's father, Bud Shuster, held the Keystone State's Ninth Congressional District seat for almost thirty years. When Bud announced his retirement in 2000, Bill immediately announced that he was running to replace him, and won.[18] As a result, for more than four decades, the Shusters have represented this region of rural Pennsylvania in Congress. And along the way, they have developed a family empire of lobbying contracts.

In January 2013, Bill Shuster returned to Washington after the Christmas break and was made chairman of the powerful House Transportation and Infrastructure Committee, a committee that his father had once chaired.[19] The influential body controls spending on roadways, automobiles, railways, mass transit, the airline industry, and pipelines.

In short, the committee moves a lot of taxpayer money around.

At the ceremony marking his appointment, Shuster brought his father, Bud, his mother, Patricia, and brother, Bob, into the chamber to introduce them to the committee members. This was unusual, even by Washington standards. "It's uncommon for family members to be introduced at these meetings," noted one industry publication.[20]

The apparently maudlin move belied an ulterior motive: both Bud and Bob were registered lobbyists for transportation interests.

Bud was running a lobbying shop called Strategic Advisors, Inc., based out of his home in Everett, Pennsylvania. Over the years he has collected millions from interests like the Association of American Railroads (AAR) and other clients. In 2012, he was a registered lobbyist for the AAR. Bill, who was chairman of the full committee, also sat on Transportation's Subcommittee on Railroads, Pipelines, and Hazardous Materials, with direct authority over the railroad industry.[21] Brother Bob was also lobbying for several transportation companies with matters before the committee. From his K Street office, he represented firms like Rajant Corporation on issues relating to "positive train control" and Canada Steamship Lines "regarding maritime transportation" issues.[22] The Chamber of Maritime Commerce, an industry trade group, was paying him to lobby on "transportation and maritime issues."[23]

The House Transportation and Infrastructure Committee holds sway on issues beyond just transportation. For example, Bill's committee oversees energy pipelines and safety regulations that pertain to them. His brother, Bob, was hired by the EQT Energy company to lobby on "pipeline safety initiatives."[24] Of particular concern to the company was a piece of legislation called H.R. 2845 Pipeline Safety, Regulatory Certainty, and Job Creation Act, which was sponsored by Shuster and went through his committee.[25] EQT was apparently happy with the work of the Shuster brothers. They were paying Bob to lobby while also giving Bill's campaigns the largest share of their political contributions over the last two elections.

Bob's lobbying firm biography notes that he "focuses his practice primarily in the areas of infrastructure projects . . . and appropriations for transportation"—areas that of course just happen to overlap with his brother's committee.[26]

Congressman Bill Shuster has a daughter named Allison Shuster who is now a lobbyist for Primerica, Inc.[27] Although her lobbying is not directly tied to the transportation committee, there are now three generations of the Shuster family circling Congressman Shuster like moons around a planet.

In 2015, yet another lobbyist joined the Shuster orbit. Shelley Rubino was not exactly family but had developed a "close private and personal relationship" with Congressman Shuster. She was, incidentally, in the transportation industry. As a vice president of Airlines for America, an airline industry association, Rubino's job was to represent airline interests to Shuster's committee. At the time of their intimate relationship, Shuster's committee was working on the overhaul of the Federal Aviation Administration (FAA) regulations. Decisions were being made about new rules for airline alliances, airline start-ups, and standards for passenger comfort. All of which would have huge implications for Ms. Rubino's employer.[28]

Technically, Shuster did not even violate any congressional ethics rules by becoming intimate with someone who was lobbying his committee. The *House Ethics Manual* is silent on having a romantic partner who is a lobbyist.[29] Sexual favors from a lobbyist have no "monetary value," so the politician is apparently free to receive them.[30]

Lobbyists for Airlines for America found yet another way into the Shuster orbit. Chris Brown, vice president for legislative and regulatory policy at Airlines for America, became Congressman Shuster's staff director on the Transportation Committee's Subcommittee on Aviation. Also, Shuster's personal chief of staff is married to the senior vice president of government relations at Airlines for America.[31]

Some lobbyists might dispute the ethical concerns of arrangements like those around Shuster. They would say that lobbying value really comes from their expertise on technical and often complex subjects, be it transportation, financial regulations, nuclear engineering, health care reform, or patent laws. These lobbyists would argue that they get paid large sums of money because of *what* they know, and how they can help legislators navigate through complex legislation, rather than *who* they know or happen to be related to.

Nothing demonstrates the speciousness of this argument more than a recent academic study published by scholars in the *American Economic Review*. They decided to test the proposition of knowledge versus relationships in lobbying by asking a very specific question: What happens when a politician switches from one powerful committee to another? Do lobbyists related or connected to the politician keep lobbying on topics related to the *old* committee, indicating that their knowledge on a subject is what matters? Or do they start lobbying on topics related to the *new* committee, demonstrating that it is their connection to a powerful politician that matters?

They found that while particular expertise was of value to the lobbying enterprise, "connections are a scarcer resource," and therefore more valuable. They write: "A lobbyist who is connected to a legislator whose committee assignment includes health care in one congress is more likely to cover defense-related issues in the next congress if the legislator he or she is connected to is reassigned to defense in the next congress." So, much of what lobbyists do appears to be predicated on whom they know and have access to, rather than what knowledge or expertise they have.[32]

Because access is key, the casual calls and texts, family meals and trips, and intangibles, like love and loyalty, uniquely position such children as politician whisperers. Get these relatives on your team, in what is essentially bribery once removed and subject to a W-2, and you shoulder your interests to the front of the line.

Chet Lott, the son of Senator Trent Lott of Mississippi, was managing Domino's Pizza franchises in Lexington, Kentucky, in 2001, when he decided to spin policy instead of pizza. He joined forces with former congressman Larry Hopkins, a Republican from Kentucky. Together, they formed Lott & Hopkins based in Washington, D.C., and Lexington, Kentucky. Soon their clients included BellSouth and a shipbuilding company. The next year, Chet launched another lobbying firm, Lott & Associates, and represented the National Thoroughbred Racing Association. By 2003, Lott graduated to the Livingston Group, founded by the former Speaker of the House, Congressman Bob Livingston, a Republican from Louisiana.

Senator Trent Lott was not the only powerful Republican to see his son settle into a lobbying empire.

Speaker of the House Dennis Hastert also saw his son's career take off. Joshua Hastert went from managing a record store called Seven Dead Arson in Illinois, to being a high-paid Washington lobbyist. "I realized that doing consulting and government relations on the Hill took up a lot less time than running a record store and brought in a lot more money," he said candidly.[33]

Lobbyists cannot legally pay a member of Congress directly. A lobbyist can hire their spouse, but the member must list it on their personal financial disclosure. Adult children are free from these rules of disclosure. Their commercial deals remain invisible. Commercial arrangements made through their children offer political figures plausible deniability. Their "fingerprints" remain off the money, while their influence still guides the flow. Why do most people want to build a family estate anyway? To pass on to one's kids.

Congressman William Jefferson, a Louisiana Democrat representing a district in New Orleans, is a cautionary tale of old-school family enrichment. His name conjures up the vivid image of getting caught with $90,000 cash in the freezer. A member of the powerful House Committee on Ways and Means, he had a global estate development plan going involving his family until the FBI raided his home in Washington, D.C., and found the money wrapped in aluminum foil in $10,000 increments and "concealed in frozen food containers."[34]

What is often forgotten in this famous case is that the frozen funds were part of a convoluted scheme involving a Kentucky Internet company named iGate. Congressman Jefferson offered to use his influence to help the fledgling company for a stake in it and a share of its profits. He managed to get iGate on the U.S. General Services Administration (GSA) list as an approved contractor, and then with every hint of new sales or investment, Jefferson sought a bigger slice of the pie.[35]

One scheme involved securing a contract in Nigeria for iGate in exchange for a larger stake in the company. Jefferson's aim was to give Nigerian vice president Atiku Abubakar $500,000 "as a motivating factor" to get iGate the contract. During his trial, Jefferson's defense team argued that he was involved simply because "he loved to help Africa." In addition, on a couple of occasions, he met with officials at the U.S. Export-Import Bank to help arrange financing for the deal.[36]

In return, investors in iGate wired tens of thousands of dollars to a company called the ANJ Group, LLC, a shell company controlled by Jefferson's family, until someone involved in the deal got suspicious and went to the FBI.[37] Jefferson was convicted and sentenced to thirteen years, and he and his family had to forfeit more than $470,000.[38]

Congressman Jefferson was trying to proxy, or offshore, his corruption by running it through entities controlled by his family. This would make it defensibly legal graft. Jefferson's mistake was being too direct and bold about it, technically crossing the line between corrupt but legal graft and illegal criminal activity—in this case, a direct bribe.

When the Kurdistan Regional Government (KRG) wanted to get the ear of a powerful member of Congress, they paid by hiring a lobbyist who was romantically involved with that congresswoman—one who would later become her husband.

Jack Einwechter was a military escort assigned to Congresswoman Loretta Sanchez. The California Democrat was a member of two powerful national security committees: the House Armed Services Committee and the House Committee on Homeland Security, giving her a unique say in American national security policy and strategy. Sanchez's husband filed for divorce and Einwechter left his wife. By 2006, Einwechter had left the military and become a lobbyist at the powerhouse firm Greenberg Traurig. Einwechter and Sanchez became romantically involved and were married in 2011. Many of his clients had business before Sanchez's committees.[39]

Over the next several years he did lobbying work for the arms manufacturer Heckler and Koch, a major Pentagon and Homeland Security contractor, Oregon Aero, a Department of Defense contractor, and a group of other active or aspiring military contractors, including Protective Group, ThingMagic, Inc., Relm Wireless, L-1 Identity Solutions, and Saab AB.[40]

Einwechter was also representing the Kurdistan Regional Government.[41] The KRG had numerous issues of concern before the

Armed Services Committee. In September 2007 as a registered foreign agent for the KRG, Einwechter arranged for Kurdish officials to meet with Congresswoman Sanchez.[42] She would also join the Kurdish-American Congressional Caucus in the House.

Loretta Sanchez and Curt Weldon may not see eye-to-eye politically—Sanchez is a liberal Democrat from California and Weldon a moderate Republican from Pennsylvania. But the two served together on both the Armed Services and Homeland Security Committees, and like Sanchez, Weldon would see those close to him profit from his position.

Weldon was first elected in 1986 and rose to become the vice-chairman of the Armed Services Committee. He had studied Russian and spoke it fluently, so it was fitting that in the chaotic decade that followed the collapse of the Soviet Union he became an important voice on U.S.-Russia relations. He served as cochairman of something called the Duma-Congress Study Group, an official body set up to foster closer relationships between the U.S. Congress and the Russian legislature (Duma). He also founded the U.S.-Former Soviet Union Energy Caucus, to promote energy ties between the two countries.[43]

Weldon's daughter Karen had studied education in college and received a master's degree in information systems. She had been working on "learning and training programs" for the Boeing Company, which had a helicopter plant adjacent to her father's congressional district, when she decided on a career change. In 2002, she set up a lobbying shop in suburban Philadelphia called Solutions North America. Her business partner in the venture was Charles P. Sexton, who had served as the finance chairman of her dad's congressional campaigns.[44]

Karen Weldon's firm quickly started collecting a profitable slate of clients, many of them from overseas. Solutions North America received a $240,000 contract from the Karić family, wealthy Serbians who had holdings in telecom, banking, and other firms. The Karić family was also linked to the war criminal Slobodan Milošević. She received that contract after her father had pushed for two members of the family, Dragomir and Bogoljub, to get entry visas from the U.S. State Department. The State Department turned the request down. According to Karen, her father had "developed a rapport" with the Karić family.[45]

Karen Weldon's firm inked a $20,000 per month contract from a Russian aerospace manufacturer that was hoping to sell drones to the U.S. military. At the time Congressman Weldon was the chairman of the subcommittee with oversight of military acquisitions. According to the U.S. Navy, Weldon pushed a program that would have led to the purchase of the company's saucer-shaped drone.[46]

Karen Weldon's firm also received a $500,000 per year contract to push "good public relations" for a Russian natural gas company called Itera International Energy. Itera had been the subject of media reports in the *Wall Street Journal* and elsewhere that alleged the firm had obtained sizable energy assets in a corrupt manner and was conducting business with entities linked to the Russian mob. The controversy was affecting the firm's ability to access capital in the United States. Congressman Weldon rounded up thirty congressional colleagues and held a dinner at the Library of Congress to honor the chairman of Itera.[47] The same day that dinner was held, Weldon introduced a resolution encouraging U.S.-Russian cooperation in developing energy resources. Two days later, he went to the floor of the House and offered a glowing assessment of Itera.[48]

Congressman Weldon was a frequent visitor to the former Soviet bloc because of his policy interest and position on national security committees. When he traveled there, his daughter would often join him. Sometimes her travel costs were paid for by Itera. On one trip father and daughter met with the president of the Republic of Georgia, who was having a dispute with Itera. Congressman Weldon reportedly helped to resolve it.[49]

When asked about his daughter's work and who her clients were, Weldon's office would parse its words carefully.

"The congressman is generally aware of his daughter's company and the work she does for several of her clients," his office told the media. "But the congressman has not discussed the specifics of Solutions North America's agreements with their clients or the nature of their representation."

Karen said that her work at the firm was primarily "legwork and project management," and that the company was "more of a business consultancy than a lobbying firm."[50]

The Weldons were the subject of a grand jury investigation, including an FBI raid on both Karen Weldon and Charles P. Sexton. Ultimately, no charges were brought in the case.[51]

The Hatches provide yet another example of a political family empire. Senator Orrin Hatch has had a long and distinguished career as a U.S. senator. Elected from the state of Utah first in 1976, he has served as the chairman of three influential committees: the Senate Health, Education, Labor, and Pension Committee, Senate Judiciary Committee, and the Senate Committee on Finance.[52] Hatch's son Scott has been a successful lobbyist for almost two decades, following in his father's wake.

Scott Hatch went to junior college and then spent four years working in the Senate Clerk's Office. His job, as he describes it, was to "take microfiches back and forth" to different committee offices. He then returned to college to finish his degree before joining the lobbying firm Parry, Romani & DeConcini. The firm was created by his father's longtime senate aide, Thomas D. Parry. Scott's job at the firm was to answer the telephone, monitor legislation, and watch congressional hearings on television. At the time he steadfastly refused to lobby. According to Romano Romani, cofounder of the firm, Scott was "very, very reluctant" to lobby. "He was worried about capitalizing on his name."[53] Whatever his initial misgivings, he soon jumped into the game.

The firm's clients included large pharmaceutical companies like Schering-Plough, which had matters sitting before Senator Hatch's committee. Of particular concern was a piece of legislation that would affect their highly profitable anti-allergy drug Claritin.[54] The firm also lined up construction and manufacturing companies like New Jersey–based GAF Corporation. The company was pushing legislation to deal with asbestos liability in the Senate Judiciary Committee, where Senator Hatch was chairman.[55]

But their biggest clients came to be the diet supplement industry, which was facing scrutiny over products like ephedra that were alleged to be linked to "severe medical problems." There was a major push in Congress to put supplements under the tighter control of the Food and Drug Administration (FDA), but the industry was vehemently opposed to this federal oversight. Producers of diet supplements were pouring money into the Parry firm where Scott was working. Meanwhile, Senator Hatch was the industry's champion. From his position in the Senate, he played a "decisive role in helping the industry fend off restrictive oversight by the Food and Drug Administration."[56]

By 2002, Scott Hatch decided to leave Parry, Romani, DeConcini & Symms (the firm's name was changed in February 2001) and open his own lobbying firm. Inhibitions about lobbying or using his family name were apparently gone now. He joined forces with Jack Martin, a former staffer with his dad who also worked at Parry Romani as a diet supplements lobbyist, and formed Walker, Martin & Hatch.[57] The firm quickly signed up major clients with matters before Senator Hatch's committee: Bayer Healthcare, Colgate-Palmolive, GlaxoSmithKline, National Nutritional Foods Association, and the Pharmaceutical Research and Manufacturers of America, among others.[58]

Senator Hatch did not see this as a problem; in fact, he said at the time that he encouraged his son to start the business. He added: "'I would have no qualms talking to Scott' about his clients. 'I wouldn't do anything for him that wasn't right.'"[59]

Scott Hatch reportedly says that he has "never personally lobbied for the supplement industry." Both he and his father say he has never lobbied the senator directly. But of course, he does not need to. That's what the partners in his firm are for. Scott Hatch is not an expert on the legislative process. He has not worked as a congressional staffer. Nor does he have any professional background in health care or in the other industries for which he lobbies. But he insists that his name does not help him in his busy lobbying and government relations career. "I don't think I get treated any different in the [congressional and government] offices," he claims. "I don't get a sense that they're saying, 'Oh, this is Sen. Hatch's son.'" The firm's success is simply a result of the efforts of "three hard-working gentlemen."[60]

In December 2007, Senator Orrin Hatch found himself having to vote "present" on an end-of-the-year spending bill. The problem? It contained a $294,000 project for a client of Scott Hatch's lobbying firm. The trouble began when a group in Riverton, Utah, wanted to get federal funds for the "Old Dome Meeting Hall Renovations" project. The request for the federal earmark was made by Senator Hatch himself. In keeping with congressional rules, Hatch had certified when he made the earmark request that "no one in his family would benefit from anything in his appropriations request."[61]

But of course, the people behind the funding request had retained his son to lobby on their behalf. That put Hatch in a dilemma. Hatch never mentioned the conflict of interest on the Senate floor, later claiming to be unaware the project was included in the appropriations request. He simply had a statement inserted into the congressional record after the fact saying that it was an "unintended and unfortunate oversight."[62]

We do not have time or space to recount every tale of "legal" corruption involving every family member of a politician who is a registered lobbyist. There are just too many, as it has become a standard conceit in the industry. Beyond the registered lobbyists, there are unfortunately even more family members who operate in the invisible twilight world of "government relations" where they do not need to register as lobbyists although they are playing largely the same influence game, and building their family empires by it.

These arrangements work because while House Ethics rules ban spouses from lobbying their own spouses' offices, they do not ban children or other family members. The children and other family members may market their access to influence as they like.[63]

Such blatant nepotism is not only tolerated on Capitol Hill, but has also developed as a highly profitable industry. It would appear that congress is more interested in building family empires than addressing the egregious ethical problem.

We close with two final examples from the senate.

Senator Dick Durbin has represented the state of Illinois in the U.S. Senate since 1997. In 2005, he became the Democratic

Party's whip.[64] Durbin's family has done well over the years as lobbyists and in securing government deals.

In 1997, shortly after her husband was elected to the U.S. Senate, Loretta Durbin set up a lobbying shop with her friend, Alice Phillips. Called Government Affairs Specialists, Inc., they have collected more than a million dollars over the years lobbying public clients in Illinois—even more from a collection of interests including pharma giant Wyeth, Chicago Title Insurance Company, and the Wirtz Corporation. The Durbins insist that he doesn't help her or her clients in Washington. She is registered as a lobbyist in Illinois, not in the nation's capital, but Loretta Durbin's clients have received federal funding that was pushed by her husband, who admits there is an "overlap" between her clients and firms receiving his help. Although Loretta Durbin insists that she does not lobby the federal government, some of her lobbying contracts suggest otherwise. A lobbying deal with the city of Naperville, Illinois, for example, says the firm would work with "state or federal government officials" on behalf of the city. Naperville has been blessed with federal grants during the time that Mrs. Durbin represented the city. Naperville's city manager, when asked if Loretta helped win these grants, answered, "I'm sure she did."[65]

Durbin's son, Paul, a lawyer who works in the energy field, also registered as a lobbyist.[66] Paul Durbin's specialty is securing public finance for a variety of infrastructure projects including renewable energy projects. "There are many federal and state programs that make developing renewable energy projects possible," he says.[67]

His father sits on the important Subcommittee on Energy and Water Development of the Senate Appropriations Committee.[68]

Senator Durbin's nephew Marty Durbin has cut a large and

profitable path lobbying for an assortment of energy interests. During his uncle's tenure in the Senate, Marty has worked as vice president of the powerful American Petroleum Institute (API), and then as president of America's Natural Gas Alliance (ANGA), before finally, in 2015, guiding ANGA to be subsumed into API.[69] According to watchdog groups, ANGA principally represented fracking companies.[70]

Fracking (hydraulic fracturing to release natural gas from shale) has been a controversial practice for energy fossil fuels, particularly among Democrats like Dick Durbin. Nephew Marty has used his position (and the lobby's money) to push their agenda. Senator Durbin has staked out a position on fracking that is much more moderate than that of some of his colleagues. In 2014 he was one of only a few Democrats to initially support the idea of expediting the export of liquid natural gas from the United States to the Ukraine.[71]

Marty Durbin has done well. In 2013, he made more than $803,000 from the gas association and related organizations and more than $250,000 from API.[72]

While the U.S. Senate requires that those who lobby its members disclose if they did so and on what bill, they are not required to report with whom they met. Ethics oversight is strangely silent on this particular detail.

Senator Patrick Leahy of Vermont is the most senior member of the U.S. Senate, having first been elected in 1975. He is an influential figure in the Senate because of his seniority, and also because he is the ranking Democrat on the U.S. Senate Committee on the Judiciary. Leahy's daughter Alicia Jackson is a lobbyist for the Motion Picture Association of America (MPAA). According to Senate records, she has lobbied on a host of issues related to the film industry, including intellectual property law. With her dad as the ranking member of the Senate Committee on the Judiciary, there is plenty of overlap. According to lobbying disclosure forms, she lobbied on the Mobile Workforce State Income Tax Simplification Act of 2015, where her father was one of the cosponsors. In 2015, she sought to influence the nomination of who would be the intellectual property enforcement coordinator for the federal government, a position created by her father's legislation in 2008. She has touched on a variety of issues that sit before her father's committee including cybersecurity, copyright law, and Internet rules.[73]

We have shown how K Street Princelings serve as a ridiculously systematic legal access point for influence. Pay the children of powerful politicians on Capitol Hill to ensure that favors follow, while these politically elite families build their empires at the expense of the American public. Next, we will see how this model replicates on a local level. City machine politicians, who once gained wealth by rigging trash contracts and steering local business to their friends, are now striking bargains with international oligarchs.

7

THE PRINCELINGS OF CHICAGO

- It is not just officials in Washington who cash in with foreign governments.
- Former Chicago mayor Richard Daley struck profitable deals with both the Chinese and the Russians.

For decades Chicago has been one of the most tainted cities in America. In 1952, the writer A. J. Liebling quoted a rare, honest alderman who dubbed the Windy City "the only completely corrupt city in America." Five decades later, in the midst of yet another city hall corruption scandal, the *Chicago Tribune* itself openly wondered if the town was now, officially, "the most corrupt city in America."[1]

So it is not surprising that the new corruption methods that are practiced at the highest levels in Washington, D.C., are also practiced in Chicago. The mayors of America's largest cities can wield an enormous amount of power, and power, in a corrupt way of thinking, makes money. Deals with foreign entities channeled through family members work as well by Lake Michigan as they do by the Potomac.

Perhaps the best way to illustrate the evolution of corruption in Chicago is by charting the changing practices of one of America's most prominent municipal families, the Daleys.

For much of the last half-century, the Daley family has run Chicago. A family of Irish immigrants from the city's South Side, two generations of the Daleys have run city hall for a combined forty-five years. Mayor Richard J. Daley ruled the city from 1955 to 1976 and became a national figure as a powerful urban boss. He was famous for toughness, and his administration's corruption scandals.[2] While he was never directly implicated in taking money, his friends, allies, and family did quite well conducting business with the city. He was once asked about city contracts being channeled to his son. "If a man can't put his arms around his sons," he answered, dismayed, "then what kind of world are we living in?"[3]

One of those sons, Richard M. Daley, grew up in his father's Chicago and learned the ways of patronage and helping friends and family. He understood the Election Day tactic "Hobo floto voto"—shuttling homeless drunks from voting booth to voting booth and paying them fifty cents each time they voted. He was certainly aware of the many financial corruption deals that swirled around city government. The son's skills in the dark arts of dirty politics earned him the nickname "Dirty Little Richie" from rival factions in the local Democratic Party.[4] Twelve and a half years after his father died in office, "Little Richie" assumed the throne and held power on the fifth floor of city hall longer than his old man. He lasted twenty-two years, electing to leave in 2011.[5]

In the past, former big-city mayors like Daley might have cashed in by joining a local law firm or company. But this former mayor of Chicago was also the brother of the White House chief of staff and a friend of President Obama. He had access to more bankable opportunities with foreign oligarchs spanning the globe.

The evolution of the Daley family's financial dealings from local to global deals reflects the development of corruption in America in general. The political class, even in our cities, may prefer foreign deals because they are harder to detect and tend to be more bankable. Faced with anticorruption legislation and transparency measures in the United States, they seek to broker their power with oligarchs and entities where they can avoid scrutiny.

For decades, Chicago has been synonymous with the very real and yet petty corruptions of big-city America: insider deals on trash contracts, favorable hiring practices, and pension schemes. And while neither Mayor Daley father nor son got rich while in office, family members did well.

During Mayor Richard M. Daley's tenure in office, Chicagoans became accustomed to financial deals and scandals involving city hall and the Daley family and friends.

In 2004, an investigation revealed that the city was spending $40 million a year for the use of privately owned trucks that were never used and sat idle. Many of these trucks were owned by felons or mob associates like Nick "the Stick" LoCoco.[6] Mayor Daley's brother, John, just happened to sell insurance to some of these trucking businesses. The mayor's friends and family made money at taxpayer expense. As one noted book describes it, "Bribes were paid to officials who gave out the trucking contracts. In the end, the trucking scandal resulted in forty-nine convictions; thirty-one of them involved city employees."[7] The mayor and family members were never charged.

In 2006, federal prosecutors charged Mayor Daley's patronage chief and top aide Robert Sorich with distributing city contracts "to well-connected companies and individuals." The same source states, "After weeks of well-publicized testimony, a judge sentenced Sorich to a forty-six-month prison term. Sorich's trial provided evidence that patronage and fixed contracts were deeply embedded practices within city government and that they probably reached beyond what prosecutors had uncovered."[8] The mayor and family members were never charged.

Corruption was systematic and widespread, but the Daleys avoided criminal prosecution. A mayor who was notorious for being hands-on repeatedly claimed ignorance of deals that favored friends and family members. When it emerged that under Mayor Daley a politically connected restaurant was given free natural gas, water, and garbage collection by the city, saving the business more than $5 million a year, the mayor's deposition was subpoenaed by attorneys. He answered "I don't recall" 139 times.[9]

As Mayor Daley ran city hall, the next generation of Princelings, son Patrick and nephew R. J. Vanecko, were making money through their city ties.

These two first garnered public attention as teenagers.

On a spring weekend in 1992, Patrick and R.J. threw a party with friends at the family's vacation home in Grand Beach, Michigan, while Patrick's parents were out of town. When some local teenagers crashed the party, Patrick fetched a wall-mounted 20-gauge, double-barreled shotgun believed to have once belonged to his late grandfather, Richard J. Daley. Vanecko took it and brandished it at the teenagers. Fortunately, the gun was unloaded, but another kid was seriously injured when he was struck in the head by a baseball bat. The local went to the hospital in critical condition.[10]

When Mayor Daley returned to Chicago, he held a press conference and sobbed openly about what had happened. "I am very disappointed, as any parent would be, after his son held a party in their home while his parents were away," he explained. "I am more deeply distressed for the welfare of the young man who was injured in this fight."[11]

Police investigated the incident and Patrick ended up pleading guilty to misdemeanor charges of furnishing alcohol to minors and disturbing the peace. He got six months' probation, fifty hours of community service, and a fine of $1,950. R.J. pleaded guilty to aiming a firearm without malice and was fined $2,235.[12] Sixteen other kids involved in the incident were charged with juvenile and adult offenses. One of Patrick's friends didn't get off so well: he was convicted of aggravated assault in the beating.[13]

In December 2004, R.J. had an even more serious run-in with the law when he was involved in a skirmish outside a bar. He knocked a man down, killing him. Not only was a judge with mayoral ties assigned to preside over the trial, stories quickly surfaced revealing that investigating officers had misrepresented and covered up evidence because of R.J.'s Daley connections. His trial played out in the headlines for weeks, and he was eventually found guilty of involuntary manslaughter and ordered to serve a sixty-day jail sentence.[14]

From their position of political privilege, Patrick and R.J. easily formed lucrative alliances with those doing business with the city. In 2003, likely through a limited liability company, they put $65,000 into acquiring a 4 percent stake in Municipal Sewer Services, a sewer service company that did business with the city of Chicago where his father was, of course, the mayor. The next year, Municipal Sewer Services was the beneficiary of a $4 million no-bid contract extension from the city. In the city of Chicago, a company is required to disclose who owns the business if they are doing business with the city. But when MSS filed its disclosure with the city, it never mentioned that Patrick and R.J. were part owners.[15]

When the *Chicago Sun-Times* exposed their ownership stake, Mayor Daley again pleaded ignorance as he had in other cases. "I did not know about his involvement in this company," Daley said about his son. "As an adult, he made that decision. It was a lapse in judgment for him to get involved with this company, I wish he hadn't done it," he said, reading from a prepared statement. He wanted people to understand "that Patrick is a very good son. I love him. And Maggie and I are very proud of him."[16]

The inspector general of the city of Chicago and the FBI started investigating in December 2007. Patrick and R.J. lawyered up; Municipal Sewer Services folded in April 2008. In January 2011, the president of the company was charged with three counts of mail fraud. Contracting as a "minority-owned" business, he illegally subcontracted the work out to nonminority businesses, including Municipal Sewer Services. Patrick and R.J. were not charged in the case.[17]

In 2004 Patrick got involved with another company called Concourse Communications that had business with the city of Chicago. The company signed a large contract to provide Wi-Fi for the city-owned airports O'Hare and Midway. The contracts were agreed to by then mayor Daley's city aviation commissioner, John Roberson, and a panel of city employees. Concourse disclosed its investors to the city but, as with Municipal Sewer, did not mention Patrick's involvement, which occurred via a venture capital fund called Cardinal Growth. On June 27, 2006, largely because of those contracts with Chicago's airports, Concourse was sold at a 33 percent profit to Boingo Wireless. Three days after the sale Patrick got a payment of $164,789. Over the next year and a half, Daley got five more payments from Concourse. His total take from the company was $708,999.[18] In June 2011, U.S. Attorney Patrick Fitzgerald filed suit on behalf of the Small Business Administration to recover $21.4 million of a $51 million small business loan that was given to the company's owner, Cardinal Growth, but had never been repaid. Cardinal Growth was liquidated.[19]

The Daley Princelings also welcomed alliances with overseas entities who were eager to do business in Chicago and needed the right connections. In the early 2000s, Patrick Daley found himself in Moscow, Russia. It is unclear precisely what he was doing there. In August 2001, he met a Russian businessman named Symon Garber. The businessman prided himself on running in Russian political circles. "It's all about who you know," he said in an interview. "It's important to be well-connected. Life is a two-way street."[20]

One key ally for Garber was a controversial Russian politician and businessman named Vladimir Slutsker, who introduced him to several high-ranking Kremlin officials.[21]

Slutsker is the largest shareholder in the Finvest Group, which owns numerous entities in Russia. Finvest has been under investigation by Russian authorities for a variety of alleged criminal activities including stock fraud. The investigations have led to the mysterious deaths of two law enforcement officials. In April 2005, General Anatoly Trofimov, the former Moscow director of the Federal Security Service (FSB), was investigating Slutsker. The FSB is the child of the KGB and is generally feared in the country because it can be vicious and aggressive. And with Russian president Vladimir Putin being a former member of the KGB, it has sympathetic figures in the highest of places. But Trofimov was brutally murdered with his wife in a car outside his flat on Klyazminskaya Street. One publication noted that "Trofimov was killed by 'a man with saltatory gait' very similar to the gait of Slutsker's chauffeur."

While organized hits are sadly not uncommon in Moscow, the killing of a former senior FSB officer in Putin's Russia is a rare occurrence. When officials investigated Trofimov's murder, Slutsker was the only person who refused to give any testimony to investigators, explaining that his parliamentary status made him immune to criminal prosecution.[22]

Months later, in September 2005, another investigator, Nazim Kaziakhmedov from the Prosecutor General's Office, was shot and killed in a Moscow restaurant. Like Trofimov, he was investigating Slutsker.[23]

Symon Garber was interested in getting into the taxi business in Chicago. And a year after he met Patrick Daley in Moscow, he launched his business. In Chicago you don't just start a taxi business. As the *Chicago Sun-Times* puts it, "In Chicago . . . , City Hall has complete control over cab companies. City officials determine the number of cabs, who can operate them, who can own them, who can drive them and how much riders pay to ride in them."[24]

Garber insists that the mayor's son played no role in launching his new venture. "Patrick didn't help me with anything," he said. "The only business deals were with a bottle of vodka." But when Garber opened his operation in Chicago, he had a very special guest on hand offering an effusive endorsement. The mayor of Chicago showed up and declared that Garber would dramatically improve taxi service in the city. "This is something we have all prayed for for many, many years."[25]

Garber's business expanded quickly. In June 2003, he controlled three hundred taxi medallions. Soon he had over eight hundred.

By 2009, his company was the largest taxi business in Chicago. The *Chicago Sun-Times* ran a headline: "Russian émigré now Chicago's Cab King; a Friend of Daley's Son."[26]

But there was a problem. In 2014, city officials began investigating claims that some of the taxis that the company was using between 2000 and 2010 were actually "salvage vehicles," meaning they were cars that had previously been in accidents and were deemed worthless and could not be used as taxis. Garber's company bought salvaged police cars and "cleaned" the titles before putting them on the streets, which is illegal in Chicago. Investigators found that more than 180 of his vehicles were salvaged. Many believe that Patrick Daley had intervened to help the company with the arrangement.[27]

The Daleys' pivot to global interests went beyond Patrick's adventures with Russian businessmen. Bill Daley, Mayor Daley's brother, saw how overseas relationships could work to one's commercial benefit. In the 2000s, while his brother was ruling Chicago, Bill was serving as chairman of the Midwest for J. P. Morgan and, beginning in 2007, was the head of corporate social responsibility for the financial giant. In that capacity, he came to play a crucial role in the firm's plans to expand its financial deal making in China. And the method he used was familiar to the ways of Chicago.[28]

Gao Jue was looking to land a highly coveted and prized position as a two-year entry-level analyst based out of J. P. Morgan in New York. But he did poorly in his job interviews. According to internal e-mails obtained by the *Wall Street Journal*, J. P. Morgan recruiter Danielle Domingue wrote colleagues that "Jue did very very poorly in interviews—some MDs said he was the worst BA candidate they had ever see [*sic*]—and we obviously had to extend him an offer." He screwed up his work visa, and "he accidentally sent a sexually explicit e-mail to a human resources employee." According to other internal J. P. Morgan e-mails, young Gao was seen as "immature, irresponsible and unreliable." But no matter. He was hired and retained by the bank anyway. A company e-mail referred to him as a "Bill Daley hire."[29]

So why did Bill Daley want to hire Gao Jue? In a move familiar to anyone following Chicago politics, Daley hired young Jue because his father was the commerce minister of China.[30]

Daley knew Chicago cronyism and corruption, and was no doubt aware of the Princeling mentality in China. He simply merged similar corrupt practices from two very different cultures. J. P. Morgan hired dozens of Gao Jues, setting off a federal investigation concerning the Foreign Corrupt Practices Act.

In 2011, just as Richard Daley was leaving the Mayor's Office, brother Bill was moving into the West Wing to serve as President Obama's chief of staff.[31] The Daleys had deep ties to Obama going back more than a dozen years. Michelle Obama was hired to work in Mayor Daley's office by a deputy chief of staff named Valerie Jarrett back when Michelle and Barack Obama were engaged. Jarrett would go on to be Obama's closest confidant in the White House.[32] Bill Daley had been mentor to Obama aide Rahm Emanuel.[33] The other members of Obama's inner circle, including David Axelrod, had "long and deep" ties to Mayor Daley.[34] In 2008, Bill Daley would play a "core role" in Obama's transition team, serving as a senior adviser.[35]

In November 2006, Bill Daley even played an important role in urging Obama to run for president. Obama drove to David Axelrod's office in River North and shared a private lunch with the J. P. Morgan executive. Obama wanted the Daley blessing on his presidential bid. "Yeah, you gotta run," Bill Daley told Obama. "Why not? What have you got to lose? Can you win? I think you can." Daley also pitched in on fund-raising, telling the young senator that he would not have any trouble raising enough money to challenge the Clintons.[36]

So, in a convergence of Princeling activity, as Bill Daley moved to the Obama White House in 2011, recently-out-of-

office brother Richard Daley was setting up a business with his son Patrick. This was no ordinary business for a former mayor and brother of the White House chief of staff, but rather a global enterprise, with a particular focus on business ties in Russia and China.[37] The business plan was peculiar because these were two disparate markets where the mayor had previously only limited dealings. What the two countries had in common was that both featured corrupt politicians, opaque business cultures, and possibly some Daley connections of reciprocity that made closing big deals easier.

The new Daley firm was called Tur Partners. A tur is a wild goat from the Caucasus region of Russia. Why the Irish Americans chose that as a name is unclear. They set up a number of subsidiaries for Tur and placed them in predictable offshore locations. Tur Partners Asia Limited was incorporated in Hong Kong, while Tur Partners Cyprus Limited and Tur Partners Eurasia Limited were incorporated in the island nation of Cyprus.[38] A popular location for Russian oligarchs to set up their business and keep them hidden, Cyprus has some of the most secretive corporate laws in the world.[39]

Joining them in these new ventures was Konstantin Koloskov, a Russian investment adviser whom they named as a principal of Tur Partners Eurasia.[40] They also struck a partnership with Mukharbek Aushev, a former Russian legislator and Lukoil executive, and named him the director of Tur Eurasia.[41] Aushev had deep ties with the Kremlin at the highest levels. No less than Vladimir Putin had granted him the country's "Order of Friendship."[42]

Former mayor Daley also joined the international advisory board of the Russian government's Russian Direct Investment Fund (RDIF), a $10 billion sovereign wealth fund run by the

Russian government.[43] RDIF was established in June 2011 to make equity investments—primarily inside Russia itself—with Kremlin money.[44] Daley, of course, had no formal background in finance or private equity. When Russian forces invaded Ukraine in April 2014, the Obama administration slapped sanctions on numerous Russian businesses. PBS's *News Hour* noted that RDIF avoided those sanctions for more than a year, citing their deep political ties in the West.

Tur Partners' connections were not only with the Russian government but also with Chinese officials. Daley had aggressively courted the Chinese government. He visited the country more than four times as mayor, and Chinese officials visited the Windy City where they found the environment extremely hospitable. "Daley promised that Chicago would be the most China-friendly city," said Chen Deming, then commerce minister of China, on one visit. Part of the reason the relationship worked was because Daley never raised any issues related to human rights.[45]

Tur Partners brought a Chinese businessman named Pin Ni into the company.[46] Ni runs the American operations of the Wanxiang Group, a large Chinese firm. Pin Ni still serves on the advisory board of Tur Partners, LLC. For good measure, Wanxiang put Richard Daley on the payroll as a consultant.[47]

As we will see in the next chapter, the controversial and secretive Wanxiang Group also has extensive assets in North Korea.

In November 2013, Chinese vice-premier Liu Yandong hired Daley's Tur Partners to handle public relations for her visit to the United States, which included an extended stop in Chicago.[48] During her visit the Chinese vice-premier met with Secretary of State John Kerry and Vice President Joe Biden.[49]

In March 2012, the Russian government announced that it

was creating a transit hub in Ulyanovsk, Russia, that would allow North Atlantic Treaty Organization (NATO) forces to bring nonlethal supplies to NATO's forces in Afghanistan.[50] At the same time a Chicago-based aircraft maintenance provider, AAR Corp., announced it was building a facility in Ulyanovsk to provide aircraft maintenance capabilities. The move was controversial among many in Russia because Ulyanovsk just happened to be the birthplace of Lenin.[51] The town is actually named after him, his birth name being V. I. Ulyanov.[52] Who was involved in the financing of the deal? Daley's Tur Partners Eurasia fund was listed as a key financier.[53] Vladimir Putin was reportedly behind the deal.[54] U.S. sanctions against Russia in light of the invasion of the Ukraine eventually put American participation in the project on hold.

In 2015, Daley's firm was granted permission by the Obama administration to solicit money from foreign investors for real estate projects in the United States. The controversial EB-5 Program, administrated by the Department of Homeland Security, allows foreign nationals to invest in projects in the United States in exchange for permanent U.S. resident status. The program has been marred by charges of corruption and favoritism, stemming from a series of EB-5 deals involving many high-profile and politically connected individuals.[55]

In the case of Tur Partners, the EB-5 seemed particularly well juiced by the Daleys because they were granted permission to raise money for a real estate project Daley himself had approved as mayor: a skyscraper that Magellan Development wanted to build on twenty-eight acres at 195 North Columbus Drive, along Lake Michigan. Tur also received permission to solicit financing to help Magellan finance another skyscraper on Lakeshore East—a proposed $900 million eighty-eight-story building dubbed Wanda

Vista Tower. That project had been approved with great fanfare by Mayor Rahm Emanuel months earlier.[56] When the *Chicago Sun-Times* drew attention to the deal, Tur bizarrely claimed they were not involved in the Vista project, even though they applied to finance one of the sites on which it is being built.

We contacted Tur Partners and former mayor Daley to ask about these deals. They did not respond to multiple queries.

We have seen how American Princelings and their families profit from these arrangements, whether their families hold power in the White House or city hall. But why do foreign actors play ball with the American Princelings? What is in it for them? Let's look at one Chinese company that has profited from its relationships with some of the players we have already seen.

8

THE HYESAN YOUTH COPPER MINE OF NORTH KOREA

- Foreign corporations with ethically sketchy business practices hire American politicians and their family members to shield themselves from scrutiny by the federal government.
- Political ties can help them overcome sanctions for even some of the worst human rights practices.

It is easy to understand why U.S. politicians like to build secret empires. Funds are hard to track and rewards enormous for their families.

But to understand why foreign corporations and governments are willing to invest in such politicians, or their families, it helps to look at these activities from a foreign entity's perspective. To that end, we will look at a foreign corporation with a troubling profile acquiring assets in the United States and around the world, including North Korea. In such a company's view and experience, investing in politicians, their families, and friends brings high-yield returns.

Wanxiang (pronounced Whon-shong) owns auto parts companies, real estate, and energy companies and has a global presence including a large footprint in the United States. It has also collected powerful political friends on both sides of the aisle, by putting them and their family members on the payroll. Along the way, it has skated past federal regulations and avoided sanctions for doing business in North Korea that have plagued other companies. It offers a powerful illustration of how politically connected firms make members of the American political class wealthy while getting special treatment in the United States.

Wanxiang is a Chinese conglomerate in the coastal city of Hangzhou, along the East China Sea that began as a "commune and brigade enterprise" (*shedui qiye* 社队企业) in 1969, a reflection of Mao's great Cultural Revolution. Lu Guanqiu, the son of farmers, built the company first by collecting scrap metal and then by maintaining and forging agricultural machinery. Next came manufacturing universal joints for cars. Wanxiang actually means "universal" in Chinese. Before his death in 2017, Lu Guanqiu was the chairman of a powerful conglomerate and was one of China's wealthiest men. Early photos show Mr. Lu meeting with foreign customers in a Mao jacket.[1] Later, he appeared in beautiful suits with American presidents, other politicians, and their family members.

Lu Guanqiu and his company succeeded thanks to his tight bond with the Chinese government. Lu began cultivating close ties to Chinese Communist Party officials in Zhejiang province, his home base, from the company's earliest days. He benefited from his "strong personal link with a particular local Party secretary named Zhu Bingshang" to keep the business going. By 1985 the Chinese Communist Party (CCP) named him an "excellent Party cadre." By 1987 he was elected to a local CCP executive post.[2]

Until his death, Lu sat on the powerful and exclusive Chinese National People's Congress, a legislative body that meets every five years in Beijing where prearranged political appointments to top party offices are put to a vote and formally sanctioned. His company also benefits from what the *Boston Globe* calls "close business ties with many government-owned enterprises, including one of the nation's largest grid operators, State Grid."[3]

Lu Weiding, Lu's only son, became president of the company in 1994. He was first admitted to the party as an alternate member of the central committee of the Communist Youth League around 2000, gaining full membership in 2007. CCP officials are littered throughout the top ranks of Wanxiang's management team. As Yang Yanle, general manager of the Work Office of the Party Committee, explained to Canada's *National Post*: "Enterprises operating on Chinese soil are under the leadership of the Chinese Communist Party. The party is an advanced organization and represents the excellent staff and citizens of the society. We will try to gather all the 'advanced members' as the core of the sub-branch of the party and make them contribute to the success of the enterprise."[4]

Lu Guanqiu credited the Chinese government for his success.[5] His wealth at the time of his death was estimated to be more than $5 billion.[6]

Wanxiang's foray into the United States began in the 1980s when they established American headquarters in Chicago. The Chinese conglomerate has vast holdings in the United States in real estate, manufacturing, energy, and other businesses. Some of these include "solar facilities that are dependent on U.S. government grants and aid."[7]

Along the way, Wanxiang has carefully chosen to do deals with America's most powerful political figures and put their friends and family members on the payroll. In 1999, just as then Texas governor George W. Bush was riding high in the Republican presidential primary, Wanxiang signed up his uncle Prescott Bush as a paid adviser to the company. The Chinese Xinhua News Agency described him as the company's "economic counselor."[8] Lu explained that "inviting Prescott Bush to be the counselor will help expand Wanxiang's operations overseas." The company's spokesman added: "He doesn't have a set of responsibilities. When we need his help, we will contact him . . . He has many friends. Even though he may not be involved in the same field as we are, he can go to his friends for help in resolving our issues."[9]

Once George W. Bush moved to 1600 Pennsylvania Avenue, Uncle Prescott stayed on the payroll.

Similarly, when Barack Obama rolled into the White House in January 2009, Wanxiang benefited from relationships with many of his Chicago colleagues and friends. Wanxiang had worked closely with Chicago mayor Richard M. Daley for years. In the spring of 2011, at the end of his tenure as mayor, Daley took a trip to China and visited the headquarters of the Wanxiang Group. He was greeted by Lu Guanqiu and as he exited the car he walked a long red carpet into the company's electric automobile division.[10]

Shortly after he left the Mayor's Office, Daley was added to the Wanxiang payroll, but has not disclosed how much he made.[11]

As noted in the previous chapter, Mayor Daley had the right connections in the Obama White House. His brother Bill served as Obama's White House chief of staff. Several senior Obama officials—and even First Lady Michelle Obama—had worked for or with him.[12]

Wanxiang's close ties in Washington were noted by the Chinese government's *China Daily*. "Wanxiang America's achievement has been associated with the US government's support and encouragement," they noted. Certainly, government officials like to encourage investments and court businesses accordingly, but Wanxiang's level of access was unusual. No less than Vice President Joe Biden invited Lu to visit several cities in the United States, including Washington, D.C., to "explore investment opportunities."[13]

Wanxiang also allied with the family of close Obama friend Penny Pritzker, Obama's finance chair and a longtime supporter who later became commerce secretary. Pritzker had family members who struck partnerships with Wanxiang, including a $1 billion real estate deal.[14]

Talk to any of the politicians or their family members forming alliances with Wanxiang and you are likely to hear that it is just like any other company. But in fact, this company in particular has been criticized by international organizations for its lack of transparency and claims of corruption. According to the corporate monitor Transparency International, the Wanxiang Group is one of the least transparent companies on the planet. In 2016 the international organization did an evaluation of one hundred companies in emerging markets—including India, Brazil, Russia, Mexico, and China. They evaluated each company for its transparency as well as the presence of anticorruption policies. Only three companies received a "zero" on a scale of "zero" to "one hundred." Wanxiang was one of them. (The other two were also Chinese companies.) Transparency International noted in a press release: "The very weak Chinese results stem from weak or nonexistent anticorruption policies and procedures, or a clear failure to disclose them in line with international practice."[15] Part of the problem was that the companies from China, including Wanxiang, disclosed "little or no financial data," the report said.[16]

Another reporter analyzing the report was more straightforward: "These enterprises did not practice any form of transparency, and can be assumed as most corrupt."[17]

Because of its tight relationship with the ruling government in Beijing, and its close ties with powerful political friends in the United States, Wanxiang has operated even in North Korea with the support of the Chinese premier while also evading U.S. sanctions. Those close ties with American politicians likely also explain how Wanxiang got Washington approval to acquire advanced U.S. technology, even though many warned that the deal severely damaged U.S. national security.

In 2009, newly elected president Obama launched a federal stimulus program that was designed to get the U.S. economy moving again. The administration pushed forward a $787 billion spending plan including infrastructure projects. Obama, a big believer in the value of alternative energy, made pouring taxpayer money into green energy companies an important component of the stimulus plan. Billions of taxpayer dollars went to wind companies, solar panel manufacturers, and biofuels. Money was also directed at companies trying to produce electric cars and new battery technologies.[18]

One major recipient was a Massachusetts-based battery company called A123. Founded in 2001, the company was developing new batteries based on small "nanoscale" materials that had been originally developed at the Massachusetts Institute of Technology.[19] The company was approved for a $249 million grant from the Department of Energy's Electric Drive Vehicle Battery and Component Manufacturing Initiative Project. The company used the money to build two battery plants in Michigan.[20] In September 2010, President Obama called the Livonia, Michigan, plant when it opened. "This is about the birth of an entire new industry in America—an industry that's going to be central to the next generation of cars," he told the assembled workers by phone. "The work you're doing will help power the American economy for years to come."[21]

It did not. The highly subsidized company struggled. Much of its business was supposed to come from electric car manufacturer Fisker, which was trying to produce fully electric cars. Fisker had been championed by the Obama administration, which gave $529 million in taxpayer loans to the company. Vice President Joe Biden publicly announced the investment of U.S. taxpayer dollars at a high-profile ceremony in Delaware. Rosemont Seneca, run in part by Hunter Biden and Christopher Heinz, owned a stake in the company. So did then senator John Kerry via Rosemont. According to financial disclosures, Kerry's share could be valued up to $1 million.

But sales were slow. By early 2012, A123 was in financial trouble. In August, Wanxiang offered to step in and buy 80 percent of the company. By October the company had declared bankruptcy.[22]

With A123 in bankruptcy, an auction was organized by the bankruptcy court. Wanxiang was bidding against Milwaukee-based Johnson Controls, a leader in mechanical and engineering systems. Wanxiang won the bid.[23]

Wanxiang's offer to purchase A123 set off alarm bells in the industry and Washington, D.C. A123 possessed industry-leading technologies that had applications not only for civilian uses but also for advanced military systems including satellites. A123 even had a few small military contracts. As former White House deputy national security advisor Mark Pfeifle put it, "In the world of advanced lithium-ion batteries, the Holy Grail has been the development of safer, faster-recycling and longer-lasting technology that will operate in extreme temperatures." A123 had developed such a technology in partnership with NASA's Jet Propulsion Laboratory, noted Pfeifle. Their technology "advances battery science by at least a decade." He warned that if the sale to Wanxiang was not stopped, "the United States may soon have to depend on one of China's wealthiest and most politically-connected Communist party leaders for access to taxpayer-funded technology that will power the next generation of space satellites and unmanned military vehicles."[24]

A bipartisan group of U.S. senators—including Democrats Dick Durbin and Debbie Stabenow, as well as Republicans John Thune and Chuck Grassley, wrote letters to senior Obama administration officials, members of the Committee on Foreign Investment in the United States (CFIUS), asking them to block Wanxiang's acquisition of A123. "The transfer of assets, technology and intellectual property, developed with American tax dollars, to a foreign company would be irresponsible," read one letter.[25]

A federal body created in 1975, CFIUS is required to review all foreign purchases of American companies, and approve or disapprove each purchase based on strategic or technological implications for America's national security. Over the years, CFIUS has reviewed energy companies, technology companies, telecoms, and military contractors.[26]

The fact was, Wanxiang was already closely linked to the White House. Lu Guanqiu had visited the Obama White House twice for meetings.[27] Wanxiang had a solar energy facility that was receiving federal government grants and loans.[28]

Former Chicago mayor Richard Daley was already on the Wanxiang payroll, too, and he extolled the company's virtues. "Wanxiang has shown itself a strong corporate citizen with a commitment to supporting local workers and developing the local economy," Daley told the *Boston Globe*.[29]

Daley was in regular contact with his brother and other senior White House officials, too.

Eventually, Weidi Lu, the daughter of Wanxiang's founder, gave $33,400 to the Democratic National Committee on May 21, 2015, according to Federal Election Commission records.[30] Her husband, Pin Ni, who heads up Wanxiang America, also gave $33,400 on the same day.[31]

Wanxiang tried to paint itself as a company independent of the Chinese government, ignoring the fact that the chairman of the company was *part* of the Chinese government. Many observers saw it otherwise. As Canada's *Financial Post* put it, Wanxiang is a "state-backed enterprise, receiving low-rate loans from the Chinese government to help it expand abroad."[32]

In addition to security concerns, the economic concern was that the company would export A123's technology to China and reduce or end manufacturing in the United States. Wanxiang and its allies allayed this concern by saying that they were not going to take the technology to China and intended to keep the technology right here in the United States.

The obvious national security issues could not be ignored, so Wanxiang tried to sidestep those concerns by focusing on A123's military contracts. It claimed no interest in them and they eventually went to a small Michigan-based defense contractor named Navitas Systems.[33] But as Pfeifle points out, this arrangement was ridiculous on its face. "Under the terms of the bankruptcy sale, the Chinese would not have access to A123 Systems' existing military contracts, but that should be of little comfort. It is the technology—the company's trade secrets and patents—that should be of concern, not its defense contracts. In fact, the company bidding for the military contracts paid only $3 million for them, compared to Wanxiang's bid of $250 million for the remainder."[34]

Jeffrey Green, executive director of the Strategic Minerals Advisory Council, declared that the arrangement was a "technical fiction." Navitas Systems would not own or control any of the technology. "Every bit of that . . . process, equipment and technology would be owned by . . . the Chinese. That's the fiction."[35]

Wanxiang's acquisition of A123 created legal problems because the technology in question fell under U.S. arms export laws, and the United States had an ongoing embargo against China. As the eight U.S. senators noted in their letter, "Not only are these products subject to U.S. export laws, but the company's research is also subject to the International Traffic in Arms Regulations (ITAR) which prohibits its dissemination to restricted foreign parties."[36]

In the end, Wanxiang's acquisition of A123 sailed through the Obama administration and won approval—with no conditions placed on the sale. It was an audacious victory. "Despite these noises from Congress, CFIUS promptly approved the acquisition in January 2013 with no refilings, no mitigations, and *no special security arrangements*" [emphasis mine], as the Peterson Institute for International Economics put it.[37]

The decision incensed many senior national security officials and former military officers. The Strategic Minerals Advisory Council, which is comprised of former U.S. senior national security and military officials, denounced the decision. By approving the deal, they lamented, the Obama administration "just allowed China to leapfrog the world in advanced batteries at the expense of American taxpayers." They went on, "The Chinese will now have direct access to US-funded and developed technology that powers our military satellites and military drones and supports our soldiers in the field."[38]

The council also rejected the platitude that Wanxiang would not share the sensitive technology with the Chinese government. "Members of the senior executive team at Wanxiang Corporation have been members of the Chinese Communist Party for decades and must be approved by the Politburo. Allowing this transaction will give the Chinese government direct access to the most cutting edge power technology or future satellite, power grid, and missile systems."[39]

Within the next year, by 2014, the Chinese government's *China Daily* was reporting that the technology and manufacturing were heading for China. "The battery, designed and developed initially by A123, is expected to be launched in a brand new Hangzhou [China] facility that is integrated as part of Wanxiang's acquisition."[40] The *Detroit Free Press* reported afterward that A123 was undergoing a "consolidation" of its Michigan manufacturing facilities. Two hundred manufacturing jobs were cut—roughly a third of the company's workforce in Michigan.[41]

In December 2013, Wanxiang bid to purchase electric car manufacturer Fisker Automotive. Despite massive government-backed loans, the company was in bankruptcy, and had an interesting list of creditors and investors. As we saw earlier, major investors included Rosemont Seneca Technology Partners. (Biden, Heinz, and Archer were also listed as personal creditors in the bankruptcy filings.) In July 2014, Wanxiang founder Lu Guanqiu arrived in the United States at the invitation of Joe Biden and met with him in Washington, D.C. Three days after Lu toured the Fisker facility in Delaware, their purchase of the company was approved.

Wanxiang's web of relationships with the political elite may also explain how the company avoids U.S. sanctions despite their commercial activities in North Korea, when even smaller companies have faced actions by Washington.

North Korea, sometimes called the Hermit Kingdom, has been ruled for decades by the Kim family. Kim Il Sung ruled the country from 1948 until his death in 1994. His son, Kim Jong Il, was supreme leader from 1994 to 2011. Kim Jong Un took the helm after his father's death in December 2011. The country, whose rulers are all brutal, all megalomaniacs, is infamous for the worst human rights conditions on the planet, abject poverty, and extreme corruption. Doing business in such a country is tough. And because of the totalitarian nature of the country, you only do business there with the approval of the Supreme Leader himself.

In 2007, Wanxiang set up a joint venture with the North Korean government to mine for copper in the country's desolate Ryanggang Province.[42] The Hyesan Youth Copper Mine, one of the largest in North Korea, is located in the Paektu Mountains on the border with China.[43] One can only wonder why "youth" is in the mine's name. According to the Korea Institute at Johns Hopkins University, Wanxiang's deal involved the Korea Mining Development Trading Corporation (KOMID). According to the U.S. Treasury Department, KOMID was "North Korea's primary arms dealer and main exporter of goods and equipment related to ballistic missiles and conventional weapons." In April 2009, KOMID was put on the UN sanctions list, which meant that Wanxiang could not do the joint venture with them.[44] So Wanxiang sealed the deal instead with the Ministry of Mining Industries of North Korea, taking a 51 percent stake.[45]

The deal was straightforward: Wanxiang would provide the capital investment to upgrade production at the mine; North Korea would provide the labor; copper would be exported to China. Both parties found the deal attractive. Upgrading and maintaining production would be cheap because pay in North Korea is one-fifth that of South Korea and a quarter of salaries paid in China.[46] At the same time, the quality of the mine was among the best in Asia. The Hyesan Youth Copper Mine had a grade of copper ore that is double that of Northeast China.[47]

The working conditions in the mine are brutal, even by North Korean standards. In 2010, it was reported that North Korean soldiers were sent to work at Hyesan by Kim Jong Il. Work conditions were so harsh that the soldiers fled and went into hiding.[48]

Wanxiang poured some $23 million into modernizing and upgrading the mine, which reportedly produces fifty thousand to seventy thousand tons of copper concentrate a year. Modernization of the mine was completed in 2011.

Wanxiang's actions in North Korea were not hidden from its friends and supporters back in the United States. Indeed, the company's English-language website reports on its "milestones" at the mine.[49]

The fact that Wanxiang was doing business in North Korea was not a secret in Washington. The U.S. State Department was aware of the deal and reported on it in official cables. The U.S. Senate Committee on Foreign Relations issued a report in 2012 on China's involvement in North Korea and noted both the existence and the importance of the mine.[50] The U.S. Geological Survey, in its annual *Minerals Yearbook: North Korea*, regularly notes Wanxiang's ownership of the Hyesan mine.[51]

Wanxiang faced some challenges. In 2009, the North Korean government tried to expropriate the mine after Wanxiang had poured money into upgrading it. Wanxiang saved the deal by making use of its close ties to Beijing. Chinese premier Wen Jiabao himself intervened and pressured the North Koreans on behalf of the company. The deal remained intact.[52]

Lu Guangiu defended doing business in North Korea on the

grounds that the country will become more liberal as a result of its interaction with Chinese companies. "Through our contact, we are certain they will become more open and more liberated," he said back in 2012.[53] There is, of course, no evidence of this happening in North Korea.

But there was one possible problem on the horizon in Washington.

By 2016, another threat emerged to Wanxiang's venture. The United Nations passed a resolution in March, and strengthened it in November, calling for sanctions against trade with North Korea, including a prohibition of the export of nonferrous metals such as copper.[54] Also in 2016, President Barack Obama followed suit when he issued an executive order declaring that companies were not allowed "to operate in any industry in the North Korean economy" including "transportation, mining, energy or financial services." Companies and individuals would face sanctions if they were found to "have sold, supplied, transferred, or purchased" anything from North Korea including "metal," which includes, of course, copper.[55]

One would expect big trouble for Wanxiang.

The *Korea Herald* reported, "The copper embargo is expected to deal a blow not only to Pyongyang, but also some Chinese mining companies. An affiliate with China's Wanxiang Group has reportedly invested more than 28 billion won ($23.8 million) in developing copper mines in the North Korean border town of Hyesan since it acquired exclusive mineral rights through a joint venture in 2007."[56]

Wanxiang might be operating the largest mine in North Korea, but the Obama administration went after other Chinese companies instead, including the Hongxiang Group. The Department of Justice charged four Chinese nationals and a trading company based in Dandong, China, with "conspiring to evade U.S. economic sanctions" on North Korea by using a front company to hide financial transactions with a North Korean bank. [57]

While there is nothing to suggest that Wanxiang was involved in a similar financial scheme, the move still mystified those who were watching Chinese activity in North Korea because those who were being prosecuted were far less involved in mining in North Korea than Wanxiang. Radio Free Asia, a news organization funded by the U.S. federal government, noted that "the Wanxiang Group is the largest importer of North Korean mineral resources, not the Hongxiang Group, which is under investigation for illegal trading with the North." They reported that "Hongxiang's imports of North Korean minerals were only a fraction of Wanxiang's imports from the North."[58]

The resulting sanctions on competitors, of course, made things *better* for Wanxiang in North Korea.

As one source told United Press International, "An investigation into companies doing business [with North Korea] should capture many large Chinese firms. The sanctions have actually given more benefits to Chinese companies." The source claimed that China feigns "compliance with sanctions and the suspension of mineral imports, but Chinese companies have been acquiring North Korea's raw materials on a 'massive scale.'" Who benefits? For one, China's Wanxiang Group, which holds an exclusive importation contract, valid until 2026, with North Korea's Hyesan Youth Copper Mine and is jointly operating the Hyejung Mining Joint Venture Co. with North Korea, which purchases all of the mineral resources in North Korea's Yanggang (Ryanggang) Province.[59]

Princeling-style corruption helps to make American politicians and their families wealthy, while foreign governments and companies get powerful favors in Washington. American laws and regulations can be bent to suit the needs of foreign entities.

This is one way that politicians use the power with which we entrust them to build their family empires. Another relatively new form of power abuse involves political "smash and grab."

9

BARACK OBAMA'S BEST FRIEND

- Barack Obama went after certain industries in the name of the public good—while his best friend positioned himself to profit.
- Barack Obama's best friend became the go-to guy when companies faced regulatory threats from the Obama administration.

Ronald and Nancy Reagan hobnobbed with Hollywood. Bill and Hillary Clinton had the Arkansas gang. George W. Bush hung out with his pals from Midland, Texas.

Each American president has brought a unique flair to the White House based on his circle of friends. For Barack Obama, that circle included a tight corps of young professionals from Chicago. Over the course of more than a decade, they played basketball, raised children, built careers, and vacationed together. In 2008, when Barack Obama was elected president, some of those friends "did" Washington with the Obamas, too.

An Associated Press writer described this small circle of friends as "regulars at President Barack Obama's side: tagging along when he accepted the Nobel Peace Prize in Norway, buying shave ice during the president's Hawaii vacation, shooting hoops in Washington, climbing a lighthouse on Martha's Vineyard off the Massachusetts coast and attending A-list White House parties."[1]

While President Obama was making landmark decisions that

adversely affected major American industries, certain close friends were making seemingly well-timed financial bets in the wake of those decisions. Executive decisions by President Obama often caused thousands to lose their jobs and tens of thousands to have their lives disrupted. Those disruptions turned out to be financial opportunities for Obama's close friends, who were positioned to profit from that political power.

Call it "smash and grab."

Barack Obama's Chicago circle was a collection of young African American professionals, some of whom are well known, such as Valerie Jarrett who joined him in the White House. But many others are little known to the American public. Highly educated and motivated peers, they gravitated to the worlds of both politics and finance, which so easily intertwine.

One of Obama's best friends is Marty Nesbitt, often referred to by mutual friends of both men as "FOB #1."[2] Or as the *Chicago Tribune* calls him, "the First Friend."[3]

Nesbitt first entered the Obama circle back in 1980 when he was recruited to play basketball at Princeton University. Already on the team was Michelle Obama's brother, Craig Robinson. After Princeton, Nesbitt moved to Chicago for business school where he met Barack playing pickup basketball games.[4]

As Obama pursued his career in law and then in politics, Nesbitt went into business and enjoyed the financial backing of a powerful benefactor, Penny Pritzker, heir to the Hyatt Hotel fortune and a member of one of the most powerful political families in Chicago. Together Nesbitt and Pritzker created a company called The Parking Spot, an offsite parking garage business with nearly forty facilities. Penny Pritzker would also go on to play a major role in Barack Obama's political rise.[5]

How close are Barack Obama and Marty Nesbitt? When

Obama ran for the U.S. Congress in 2000 against Congressman Bobby Rush, he tapped Marty Nesbitt to be his campaign chairman. Obama lost. When Obama ran for the state senate, Nesbitt was a key fund-raiser. When he ran for the U.S. Senate, Nesbitt was again on the finance committee. When Obama ran for president in 2008, he served as the campaign's treasurer.

Barack Obama is the godfather to Nesbitt's son, and Nesbitt's wife, a physician, delivered the Obamas' two daughters. Along the way, they have taken vacations together and played golf and countless games of basketball.[6]

Once Barack Obama became president, Nesbitt was a regular presence at the White House. When the Obamas vacationed in Hawaii or in Martha's Vineyard, the Nesbitts were always there. Marty Nesbitt attended state dinners and barbecues on the White House South Lawn. They were in regular contact. In 2014, when Barack Obama was contemplating life after the White House, he asked Marty Nesbitt to serve as the chairman of the Obama Foundation.

During Obama's tenure in office, Nesbitt had the direct access to the president that you might expect of an intimate friend. As he explained in one interview, "I don't have to call to schedule a time during the work day to talk to him, because I'll be like, 'I'm going to be here this weekend, we're going to go play golf, then I'll have time to run a couple of things by you.'"

Nesbitt described their relationship this way to the *Washington Post*:

> "Every now and then, someone will send me on a mission because he needs to be told something that only I can tell him," Nesbitt said. "I'll say, 'Hey, you know, they think you should be doing this.' And Obama will respond, 'You are just so trans-

parent.'" Nesbitt added: "When he knows we're just trying to cheer him up or pump him up, he'll just start laughing."[7]

In early 2013, as Barack Obama was celebrating his reelection, Marty Nesbitt was preparing to launch a new venture: a private equity investment firm called Vistria. The word is a combination of the Latin words "power" and "three."[8] Nesbitt was launching the firm with another former basketball player named Harreld "Kip" Kirkpatrick III. Kirkpatrick had run for state treasurer in Illinois and was well-connected in the Obama network. Kirkpatrick was also helping to build a company called United Shore Financial Services, formerly Shore Mortgage. At Shore, Kirkpatrick left the nuts and bolts of running the company to others and spent much of his time in Washington, "calling on politicians" or "meeting with executives from the federal government backed Fannie Mae and Freddie Mac, the folks who buy the loans" that his company originated.[9]

Vistria was launched with money from Penny Pritzker, partly through an entity called the Pritzker Traubert Family Foundation.[10] They also secured funds from state government employee pension funds, including the Illinois Municipal Retirement Fund and the New York State Retirement Fund.[11] In both instances, these state retirement funds advanced the money through programs designed to boost investment business with minority-controlled investment funds.

The investment business model at Vistria was unusual. While many investment funds shy away from investment deals involving highly regulated industries because of the red tape, Vistria was actually focused on those sectors. As Marty Nesbitt told his alumni magazine: "We will look at acquiring companies that are highly regulated . . . That kind of search leads to sectors such as education, healthcare, and financial services."[12] On another occasion, he described Vistria's focus on companies that are at "the nexus of the public and private sectors."[13]

Investing in highly regulated industries made sense, given that Nesbitt was best friends with the Regulator in Chief. An investment fund website noted about Vistria, "Both [Nesbitt and Kirkpatrick] are long-time politicos in Illinois, and they are targeting companies operating in highly-regulated industries like education, healthcare and financial services that will benefit from their expertise, and presumably their connections."[14]

Marty Nesbitt launched Vistria in sync with his friend President Obama's reelection. According to corporate records, Nesbitt filed for a trademark on the name Vistria on December 11, 2012, just a month after his friend was reelected. Ten days later, he was off for the annual Christmas vacation to Hawaii that his family took together with the Obamas. From December 21 to January 5, 2013, they golfed, dined, and walked on the beach together.[15] In the month Vistria became public, on March 30, 2013, Nesbitt attended an NCAA basketball tournament game with Obama.[16]

To help Vistria invest in highly regulated sectors, Nesbitt and Kirkpatrick brought on board several regulators and insiders from Obama's administration. One was Tony Miller, who had been appointed, in July 2009, deputy secretary of education in the Obama administration and had played a central role in the development of a wide array of policies and regulations at the Department of Education. Miller left the DOE in July 2013 to join Vistria shortly after it opened its doors.[17] During this time, Nesbitt and Obama remained close. They golfed together at Andrews Air Force Base and then stayed together at Camp David on August 3 and 4, 2013.[18]

Another early hire was Jon Samuels, who was the deputy assistant to President Obama for legislative affairs.[19] On his LinkedIn page, Samuels describes his regulatory work in the Obama White House: "Strong focus on Dodd-Frank legislation. Significant work to pass the Affordable Care Act."

Also on his LinkedIn page, Samuels says of his duties at Vistria: "Private Equity, Public/Regulatory Policy, Government Relations, Business Development, Strategy."[20]

Fortune magazine noted Samuels's hiring and the fact that he "doesn't appear to have any experience working in the financial services industry. Rather, Samuels has made his career in politics."[21]

The industries that Vistria targeted to buy were the same industries that Nesbitt's friend, President Obama, was targeting with a series of government actions. Vistria's new hires, Miller and Samuels, had been instrumental in these massive government actions, which included everything from new regulations, legal actions, and legislative threats.[22] A curious pattern began to emerge. Obama and his administration would attack industries with government power, which led to substantially lower valuations for these companies. Nesbitt and Vistria, or others close to Obama, could then acquire those assets for pennies on the dollar.

The Obama administration had several industries in their crosshairs, deeming them destructive to the environment or exploitative of people. Industries such as coal mining, offshore energy companies, cash advance companies, and for-profit colleges became targets for litigation, regulatory squeeze, or denial of access to government services or funds. A circle of investors including Vistria and others linked to Obama would consistently purchase companies in these sectors once their valuations dropped under the government onslaught.

One of the Obama administration's earliest and most visible targets was the for-profit higher education industry. While millions of young adults attend public or private nonprofit colleges and universities to continue their education after high school, millions also opt to attend for-profit schools. For-profits often allow flexibility in scheduling and many trade skill programs. These for-profit schools include well-known names like the University of Phoenix, ITT Technical Institute, and DeVry University.[23]

President Obama concluded in 2013 that some of these schools victimized students. "They've been preyed upon very badly by some of these for-profit institutions." He said, "Their credit is ruined, and the for-profit institution is making out like a bandit." Military veterans, he declared, were often "manipulated" by these schools, which recognized "there was a whole bunch of money that the federal government was committed to making sure that our veterans got a good education, and they started advertising to these young people, signing them up, getting them to take a bunch of loans, but they weren't delivering a good product."[24]

Defenders of for-profit colleges would say that they are viable gateways to opportunity for many who might not otherwise have access to career development. Their students are often older students, half of whom work full-time. A third are raising their own kids while attending school. Many students like the schools because they provide more convenient locations than traditional schools. Their record for graduation can be superior to nonprofit schools. As one researcher notes, "For all their widely publicized shortcomings, for-profits turn out to be better than community colleges at graduating students from 2 year programs."[25]

The pros and cons, or reasoned logic, was less important than the federal government's leverage: in any given year, these for-profit schools received as much as $32 billion in federal student aid.[26]

Government leverage from the Obama administration began in 2011 when the Department of Education announced implementation of a so-called gainful employment rule, which would require for-profit schools to track the performance of their graduates in the job marketplace. Those programs that did not deliver good outcomes would be cut off from federal grants and federally backed student loans.[27]

The DOE meetings to craft regulations included names who were now connected with Vistria. According to the Department of Education, Tony Miller, as the deputy secretary of education, participated in those private meetings with for-profit schools. So did Arne Duncan, the secretary of education. Both would eventually leave DOE for Vistria—Miller taking a position as chief operating officer, Duncan maintaining an office there.[28]

While the new regulations had their plausible merits, it quickly became clear that there was money to be made by investors specially attuned to DOE regulatory changes. E-mails indicate that senior Department of Education officials were communicating with Wall Street investors about the new regulatory rules.[29] Wall Street hedge funds were buying and selling the stock of for-profit colleges and universities, so details on what the rules said, and how and when they would be implemented, could make or cost traders a lot of money.

The left-leaning Citizens for Responsibility and Ethics in Washington (CREW) obtained e-mails that indicated senior Department of Education officials were actively communicating with hedge fund investors that were "shorting" (betting that prices would go down) for-profit college stocks based on the new rules. CREW found that both the deputy undersecretary of education and the budget development staff director "carried out a planned leak of the proposed gainful employment regulations to a number of outside individuals and groups in advance of the regulations' public release. This effort started with an e-mail from hedge fund short-seller Steven Eisman." CREW argued that:

> high-level Education officials involved in the agency rulemaking process not only knew of the efforts of certain hedge fund managers to influence the regulatory outcome, but may themselves have colluded with those individuals to protect the short-sellers' financial interests. They also document a plan by high-level Education officials to leak the contents of the gainful employment regulations in advance of their public issuance.[30]

Two U.S. senators sent a letter to the education inspector general asking for an investigation. "The Department may have leaked the proposed regulations to parties supporting the Administration's position and investors who stand to benefit from the failure of the proprietary school sector."[31]

In the U.S. Senate, a diverse group of officials including Senators Joe Lieberman of Connecticut and Michael Enzi of Wyoming expressed deep concerns about short sellers trying to game the market based on what the Department of Education would do.[32]

The gainful employment rule by the Department of Education was only the beginning of regulatory fire. Soon came "a broader series of crackdowns on the industry by agencies including the Consumer Financial Protection Bureau, the Federal Trade Commission and the Securities and Exchange Commission" against for-profit schools. The industry felt they were clearly under the gun.[33]

"We've come to expect these unjust assaults," said Gene Feichtner, president and chief executive officer of the ITT Technical Institute. "Let there be no presumption here that we believe we'll be treated fairly."[34]

Particularly hard hit by the Obama push against for-profit colleges was the Apollo Education Group, which operated the University of Phoenix. The school was well-known because it had satellite campuses around the country, advertised regularly, and even sponsored a football stadium for the NFL's Arizona Cardinals. Beyond grappling with the new gainful employment rule, Apollo was forced to play defense by the feds.

In July 2015 the Federal Trade Commission announced an investigation into Apollo "regarding potential deceptive advertising, sale or marketing of its services to students."[35] By October, the Department of Defense put the University of Phoenix on probation. On October 7, 2015, Pentagon official Dawn Bilodeau sent a letter to the University of Phoenix imposing the suspension and specifically cited the large military bases at Fort Bragg, Fort Carson, Fort Hood, and Fort Campbell as being affected.[36] The school had been receiving $2 billion to almost $4 billion a year in taxpayer funds, largely because so many soldiers, sailors, and marines were attending the school.[37] But that income stream was now under threat.

The move against for-profit universities was a dramatic blow to companies—but also to veterans. Unlike students with federal student loans, the GI Bill benefits vanish when the thirty-six-month tuition period runs out. Furthermore, the GI Bill includes housing for school, which means that vets on the GI Bill attending for-profit schools were losing their housing payments. Some actually became homeless as a result.[38]

The DOD action against the University of Phoenix was swift and surprising. The school appears to be one of a handful of schools that were singled out. About fifteen schools had committed similar offenses, but the Pentagon only suspended four of them. The four suspended were all for-profit schools and included the University of Phoenix. In 2016, Peter Levine, acting undersecretary of defense, told a congressional panel that the matter was handled poorly. "I think the process was crappy," he said.[39]

But the damage was done.

It was a devastating blow to the company. The stock price declined from $11.29 a share in October 2015 to $6.38 by January 2016. The price in January 2016 was off over 90 percent from where the company's stock was in January 2009 when Obama took office.[40]

The story might have ended here, with a company's decline and eventual collapse under government pressure. But then suddenly, riding to the rescue, was a group of investors, among whom was President Obama's best friend.

That group included the Wall Street firm Apollo Global Management (not previously affiliated with Apollo Education Group), an Arizona-based investment firm called Najafi Companies, and Marty Nesbitt's Vistria Group. Both Apollo Global Management and Najafi Companies are large players in the private equity field. They have large investments around the world. Vistria was a small player in comparison. Only a couple of years old, it was clearly the smallest of the three. What did they bring to the table? Apparently, Princeling connections with executive-branch approval insurance.

Before the sale was finalized, the deal required the approval of the Department of Education where, of course, Vistria's Miller used to be a senior official. There was also the all-important matter of getting the federal student money flowing again to the University of Phoenix from the Obama administration.

Vistria and their partners hired D.C. power attorney Jamie Gorelick, who served as a deputy attorney general in the Clinton administration, to lobby the Pentagon to lift the suspension on federal money, and she did so successfully in January 2016.[41]

As Nesbitt and his team of investors awaited Obama administration approval for the deal, he attended White House events, showed up at an intimate Obama birthday party celebration, golfed with him on Martha's Vineyard, attended fund-raisers, and they played basketball together.[42]

By December 8, 2016, the Obama administration approved the sale of the Apollo Education Group, but there were several conditions. The Department of Education required the company to submit a letter of credit valued at 25 percent of the federal funding they were expecting to get via student loans and grants. That amounted to about $386 million. It was a condition for approving the sale.[43] Vistria and its coinvestors claimed that these demands were excessive and would put the deal at risk. Less than two weeks later, on December 20, just weeks before Obama was to leave office, the Department of Education lowered that credit requirement from 25 percent to 10 percent. The other 15 percent could be put in escrow.[44]

On February 8, 2016, the holding company for the University of Phoenix (Apollo Education Group) announced that it was being sold. The transaction was valued at $1.1 billion.[45] Apollo Global Management, Najafi, and Vistria bought it for just $10 a share, or $1.14 billion.[46] It is worth restating that the company, before the regulatory onslaught during Obama's tenure, had been worth almost nine times that price. With the sale, Tony Miller became the chairman of the board.[47] Miller, of course, had been the number two official in the Department of Education when the campaign against for-profit colleges had been launched, and a participant in those pivotal meetings. Arne Duncan set up an office at Vistria's headquarters in Chicago, overlooking Chicago's Millennium Park.[48]

So, Obama's best friend and senior officials in his administration gained control of a for-profit college in a bargain deal. Better yet for Vistria, competitors were smashed and were now out of business. Corinthian Colleges, Inc., was run out of business. Arne

Duncan, then still the secretary of education, explained in 2015 that he was "thrilled to be able to close down Corinthian." ITT was essentially shut down, eliminating almost all of their eight thousand jobs and leaving tens of thousands of students stranded without a school or a degree.[49]

Those in the world of education took notice. "I think every way you look at this transaction is questionable and suspicious," said Diane Jones, former assistant secretary for postsecondary education at the Department of Education under President George W. Bush. Jones allowed that the DOE's changing terms after a public recommendation is highly unusual.[50]

"There is at least a taste of unseemliness involved in this," said Mark Schneider, a senior education official during George W. Bush's presidency. "They regulate it. They drive the price down . . . They are buying it for pennies on the dollar."[51]

Congresswoman Virginia Foxx, who heads the House Committee on Education and the Workforce's Subcommittee on Higher Education, said, "It's ironic that a former senior official at the Department of Education—an agency that has intentionally targeted and sought to dismantle the for-profit college industry—would now take the reins at the country's largest for-profit college."[52]

I contacted Nesbitt, Vistria, and others about this deal but they failed to respond. Former education secretary Duncan got back to me and said that there was no favoritism in this deal.

Nesbitt was not the only Princeling ready to grab from the spoils in the for-profit college space.

One investment fund that was aggressively buying and selling stocks in for-profit colleges as valuations fluctuated in the wake of Obama's push was called Ariel Investments. It is run by John Rogers, who, like Marty Nesbitt, is a close Obama friend. He is also "best friends" with Michelle Obama's brother, Craig Robinson.[53] Rogers was one of Barack Obama's earliest financial supporters when he ran for the state senate back in Illinois. Rogers was also part of the Chicago circle who traveled and socialized with Obama. By 2016 he had already donated between $500,000 to $750,000 to the Obama Foundation.

Throughout the Obama administration, Ariel was actively buying and selling shares in a number of for-profit schools. In late 2010, Ariel held $121 million worth of shares in DeVry, which operates training schools around the country. They also held a small $4.63 million in Apollo Education.[54] But by early 2011, their position in DeVry had increased to $215 million (3.91 million shares), and their stake in Apollo Education had jumped to $33 million (802,945 shares).[55] Their investment in DeVry represented their second-largest position at the fund. By mid-2011 they had cut their position in DeVry by close to 25 percent. These sales of stock coincide with when Senator Charles Grassley and others in Washington began looking at whether the Department of Education officials were leaking confidential government information to hedge fund managers.[56]

Ariel continued to trade in the stocks of for-profit colleges throughout the Obama years.[57] After Secretary of Education Arne Duncan resigned his post effective January 1, 2016, besides establishing an office at Vistria, he also joined the board of directors of Ariel Investments.[58]

Another of the "highly regulated" sectors Marty Nesbitt's Vistria moved into following in the wake of an Obama administration regulatory push was the financial industry. Recall that Vistria had hired Jon Samuels, an aide to President Obama who had helped push through the financial reform bill known as Dodd-Frank.

The law, which was signed by Barack Obama in July 2010, created a new cluster of regulations that affected virtually every aspect of the U.S. financial industry and gave the federal government broad regulatory powers over financial industries that had previously been regulated mostly at the state level.[59] One of the hardest hit was the cash advance industry, which provided short-term loans to individuals and businesses.

Plenty of abuses had taken place in the industry, but many have argued that these lenders play an important role in the financial lives of millions of Americans, whether they be business owners or individuals. According to the 2011 FDIC (Federal Deposit Insurance Corporation) National Survey of Unbanked and Underbanked Households, more than thirty-three million households were unbanked (no accounts whatsoever) or underbanked (accessed alternative financial services like payday lending), primarily because banks don't issue credit to people with high credit risk. So cash advance businesses played an important role in the economy by providing cash to people who might need it for short periods of time but who couldn't qualify for a traditional loan.[60] Economists at the Federal Reserve Bank of New York also challenged the notion that cash advance companies were that bad, arguing that most of the critique was exaggerated. "Many elements of the payday lending critique—their 'unconscionable' and 'spiraling' fees and their 'targeting' of minorities—don't hold up under scrutiny and the weight of evidence," they wrote.[61]

Regardless of where one falls on the issue, there can be no debate about what followed. In July 2011, as a result of the new law, the Consumer Financial Protection Bureau (CFPB) opened its doors. The new federal government regulatory agency had a dual mission: educating Americans on financial matters and also investigating what it regarded as "unfair" lending practices, particularly by the consumer finance industry. By December 2012 the Obama Department of Justice was teaming up with the CFPB team to crack down on what it deemed to be financial crimes. "The fair lending enforcement work of the CFPB and the Justice Department is part of efforts underway by the Financial Fraud Enforcement Task Force (FFETF) which was created to wage an aggressive, coordinated, and proactive effort to investigate and prosecute financial crimes. The CFPB and the department's Civil Rights Division are among the co-chairs of the FFETF's Non-Discrimination Working Group."[62] Beginning in 2013, the Department of Justice and the CFPB would begin to aggressively investigate, audit, and generally harass payday and small loan lenders.

The creation of the new government agency almost immediately caused waves in the industry. According to the 2012 SEC filings for First Cash Financial Services, Inc., the CFPB could "make the continuance of all or part of the Company's current U.S. business impractical or unprofitable."[63] Other lenders shared similar warnings. Within the cash advance industry, it was widely believed that the new regulations would lead to heavy burdens for smaller lenders. A report done by The Heritage Foundation estimated that the new regulations would throw many lenders out of business, stating that "the new regulatory strictures will . . . reduce consumers' choices of financial products and services."[64] This would then winnow out the competition and create opportunities for other firms to acquire them.[65]

Operation Choke Point was launched by the U.S. Department of Justice in 2013 and targeted banks and the business they do with payday lenders, payment processors, and other financial companies they believed to be at higher risk for fraud and money laundering. The government was reportedly pressuring the financial industry to cut off the access to banking services including access to capital or loans. Banks terminated their accounts with payday lenders because of the federal scrutiny.[66] A lawsuit claimed numerous check cashers and payday lenders had their relationships terminated because of the policy.[67] As one executive in the industry put it to an industry publication, in 2016, some lenders were being "squeezed out of the business by regulations like Operation Chokepoint."[68] At a House Financial Services Committee hearing on April 8, 2014, it was argued that Operation Choke Point was hurting nonbank financial service providers by reducing their access to the banking system.[69]

The effect on the industry was clear: the number of financial institutions offering these loans began to shrink. Regular banks, which actually offered cash advances more than one might think, got out of the business in "response to guidance from other regulators." There began what the *Washington Post* called "serious retrenchment" in the industry.[70]

The closing of some businesses, of course, created opportunities for other lenders who were able to adapt to the new rules.

One of those businesses that thrived in the wake of the new rules was a small lender based out of California called ForwardLine Financial. The company had some well-connected friends. In October 2015, ForwardLine Financial announced that it had a new chairman: none other than President Obama's best friend, Marty Nesbitt. Also joining the board was Vistria associate Michael Castleforte. ForwardLine was founded in 2003 and was designed to provide alternative financing solutions to small businesses, and as a provider of small business merchant loans and cash advances. ForwardLine uses "non-traditional credit algorithms to finance 98% of U.S. businesses that banks consider too small and too risky for a business loan."[71]

ForwardLine's business interests were directly affected by Dodd-Frank legislation. As ForwardLine's competitors were choked out of the finance industry by mechanisms like the CFPB and the DOJ's Operation Choke Point, ForwardLine actually saw its business grow.[72]

One of ForwardLine's competitors, Advance America Cash Advance Centers, Inc., sued the federal government, charging that they were being targeted in such a way that they could not get banks to work with them, alleging they were the object of "regulatory retaliation."[73]

ForwardLine has critics who argue that the company engages in some of the industry's worst practices. They accuse the company of charging excessive interest rates. One man who got a $33,000 business loan claims that he was given a one-month loan at a 30 percent interest rate.[74] Others, including consumer evaluators, complain that the company is not transparent about interest rates and fees.[75]

Princeling Marty Nesbitt also enjoyed other boons arguably based on his blood-like ties to the president, in the form of corporate board appointments from major corporations that faced serious regulatory challenges from the Obama administration. He joined the boards of several companies in industries for which he had no background or history.

From the earliest days, the railway industry was feeling the regulatory heat of the Obama administration. Back in 1980, Congress sent, and President Jimmy Carter signed, the Staggers Rail Act, which "largely but not completely deregulated the U.S. freight rail industry."[76] That had essentially been the consensus point of American politics until the Obama administration announced its intentions to reregulate the industry.

The Obama administration was consistently pushing for new taxes and regulations on the industry. With Obama's reelection in 2012, the railroad industry was on notice that they faced the prospects of "re-regulation" after more than thirty years.[77] As Norfolk Southern Railroad put it in its 2012 Annual Report (10-K) filing with the Securities and Exchange Commission: "Efforts have been made over the past several years to re-subject the rail industry to increased federal economic regulation, and such efforts are expected to continue in 2013 . . . Accordingly, we will continue to oppose efforts to reimpose increased economic regulation."[78]

The Obama administration regularly proposed new freight rail regulations to congress.[79] Further, the Obama administration's Surface Transportation Board (STB) became a "wholly independent federal agency" during his tenure. Along with the Federal Railroad Administration (FRA), the STB had widespread regulatory control over railroads' environmental, safety, and security practices. They determine rates, routes, fuel surcharges, conditions of service, and the extension or abandonment of rail lines. Together, they also had jurisdiction over consolidations and mergers among rail common carriers, and regulated certain track and mechanical equipment standards.[80]

Weeks after President Obama was reelected and sworn in for his second term, Norfolk Southern announced that Marty Nesbitt was joining their corporate board. Nesbitt had no background in railroads or transportation unless you count airport parking garages as transportation.[81] His 2015 compensation was $278,937.[82]

Another company facing aggressive regulatory problems with the Obama administration was American Airlines. Just as Marty Nesbitt was joining the Norfolk Southern board, American Airlines was in serious financial trouble. Facing bankruptcy and even possible financial collapse, American announced a merger with US Airways. But the Obama Justice Department immediately filed suit to stop it on the grounds that the merger was anticompetitive. Members of Congress soon stepped in to offer their support for the merger. Sixty-five Democratic congressmen and congresswomen signed a letter in support of American.[83]

The letter, addressed to President Barack Obama, stated:

> We believe DOJ's legal challenge puts at risk the future economic security of our constituents, tens-of-thousands of unionized workers at both airlines, and the economic well-being of communities that we represent . . . We are concerned that the DOJ's lawsuit creates an atmosphere of uncertainty for our respective congressional districts and constituents. While we share your concern regarding any potential impact on consumers as consolidation in any industry is contemplated, we believe that DOJ's concerns as outlined in the complaint filed last month are not an adequate representation of all of the facts.[84]

In light of the congressional pushback, the Obama administration backed down from blocking the merger.[85] But American Airlines' struggle with the Obama White House continued. The company had serious pension problems and was embroiled in a struggle with the federal government's Pension Benefit Guaranty Corporation.[86] There was also a federal investigation launched by the Obama administration into alleged price gouging. The investigation was announced in July 2015.[87] Additionally, American Airlines was pressing the Obama administration to renegotiate the so-called Open Skies Agreement, which they argued created unfair competition.[88]

By October 2015, *US News* was reporting that American, as well as two other big carriers, were lobbying to "shield us from competition and roll back consumer protections."[89] American Airlines "emerged as a leading voice in the bumpy campaign to persuade the White House to intervene in what it calls unfair competition from foreign rivals." They argued that the governments in certain countries were funneling billions of dollars into subsidies for their airlines, creating an unfair advantage.[90] In September 2015 the CEO of American Airlines met with secretaries of transportation, state, and commerce to discuss "complaints against Emirates, Etihad, and Qatar airlines."[91]

On October 17, 2015, the merger between American Airlines and US Airways took place. Less than a month later, American Airlines appointed Marty Nesbitt to their board of directors.[92] Since Nesbitt had no background in the airline industry, his appointment must have been for some other reason. In 2016, Nesbitt cleared $395,704 for his board membership.

American Airlines has a history of seeking such political alliances. When Senator Tom Daschle in 2001 was the most powerful Democrat in the Senate, his wife was a highly paid lobbyist for American Airlines.[93]

Today, as chairman of the Obama Foundation, Marty Nesbitt is, as *Politico* puts it, "the man building Barack Obama's future."[94] From an empire-building perspective, this payback makes sense, given that Obama has already helped Nesbitt build his legacy.

10

MORE SMASHING AND GRABBING

- The Obama administration saw the rise of "smash and grab" as a new form of cronyism.
- As Barack Obama went to war with the coal industry, some of his closest financial backers—Tom Steyer, George Soros, and others—positioned themselves to profit.

Barack Obama walked into the ornate offices of the *San Francisco Chronicle* and took his seat behind a microphone. Amicably meeting with a group of the paper's journalists, he spent the next hour talking about his vision for the country. It was January 2008 and the Democratic Party presidential primary was just heating up. On this particular day, he covered a wide variety of subjects, including one he was particularly passionate about—the environment. As the journalists listened, he laid out his concerns about the warming of the planet, and how the fossil fuel industry—oil, gas, and coal—was largely responsible for it. He was promising not only to talk about it but do something. "If somebody wants to build a coal-powered plant, they can; it's just that it will bankrupt them because they will be charged a huge sum for all that greenhouse gas that's being emitted . . . Even regardless of what I say about whether coal is good or bad. Because I'm capping greenhouse gases, coal power plants, you know, natural gas, you name it—whatever the plants were, whatever the industry was, they would have to, uh, retrofit their operations."[1]

This was not new for Obama. On October 8, 2007, Obama had promised before a crowd in New Hampshire that he would use "whatever tools are necessary to stop new dirty coal plants from being built in America—including a ban on new traditional coal facilities." Then he criticized the oil industry for getting favorable tax breaks and contributing to global warming. "When we let these companies off the hook; when we tell them they don't have to build fuel-efficient cars or transition to renewable fuels, it may boost their short-term profits, but it is killing their long-term chances for survival and threatening too many American jobs. The global market is already moving away from fossil fuels."[2]

Some in coal country continued to hope that Obama would prove to be a friend. Shortly after the 2008 election, the American Coalition for Clean Coal Electricity (ACCCE), a coal industry group, placed an ad of video excerpts from a campaign speech Obama had given in September that year in Lebanon, Virginia. In the excerpts, Obama said that America could be "energy independent" with "clean coal technology." He explained, "This is America. We figured out how to put a man on the moon in 10 years. You can't tell me we can't figure out how to burn coal that we mine right here in the United States of America and make it work." The advertisement ran for several months.[3]

Coal-state Democrats agreed with the ACCCE hope that Obama would prove to be a coal supporter. During the 2008 campaign, they argued that Obama would be a friend of coal producers and workers. Congressman Nick Rahall of West Virginia declared that Obama would be "better for the industry than John McCain."[4] Their hope would eventually go up in smoke.

During his eight years as president, Barack Obama went after numerous industries, charging them with being damaging or dangerous to the American way of life. We saw in the last chapter how his administration worked to undermine the for-profit education system on the grounds that it was ineffective and exploitative. We also saw how after he had driven those companies largely into the ground, his friends swooped in for strategic purchases for pennies on the dollar. Obama pushed what seemed an all political agenda (smash), while just happening to make his wealthy friends even wealthier (grab).

Perhaps the industry that Obama and his administration most persistently targeted was the fossil fuel industry. Petroleum producers and the coal industry faced a steady stream of criticism, regulation, and restrictions on their commercial activities during his tenure. They faced repeated attempts to tax them heavily. In many instances, Obama's closest political allies and financial backers were poised to buy up coal and oil companies or shares once these assets lost value in the wake of his activities.

Obama had many ambitions when he came into the White House. Health care reform. A stimulus program to create jobs. But perhaps none lasted longer throughout his eight years in office than his ambition to transform the American energy industry. Obama made clear that he wanted to reduce the power and size of the fossil fuel industry. He considered it a dirty relic of the past that was substantially harming the environment. He also wanted to expand and bolster the alternative energy industry: solar, wind, biofuels, and thermal. On Earth Day, 2009, in Newton, Iowa, he voiced his resolve. "Everybody has known that we had to do something but nobody wanted to actually go ahead and do it because it's hard," he told the crowd. "I reject that argument."[5]

Months earlier, after just weeks in office, Obama unfurled his ten-year budget, seeking to eliminate "oil and gas company" tax breaks and adding a "new excise tax on offshore oil and gas production in the Gulf of Mexico."[6] Good-bye George W. Bush administration, which had been generally supportive of the energy industry, there was a new sheriff in town and he was not friendly with the traditional energy industry. Indeed, *Entrepreneur* magazine pondered the question, "Can Exxon Mobile survive Barack Obama?"[7]

On April 20, 2010, disaster struck in the Gulf of Mexico when the Deepwater Horizon, an offshore oil rig operating deep on the ocean floor, exploded, causing a major oil spill.[8]

The ramifications of the disaster included emotional justification for a regulatory war. On April 30, President Barack Obama ordered the federal government to hold the issuing of new offshore drilling leases until a review determined whether more safety systems were needed and authorized teams to investigate all oil rigs in the Gulf, in addition to the investigation of the disaster.[9]

Obama explained to the American people that he was "frustrated and angry" about the BP oil spill and that his daughter Malia was upset, too. "And it's not just me, by the way," he said. "When I woke up this morning and I'm shaving and Malia knocks on my bathroom door and she peeks in her head and she says, 'Did you plug the hole yet, Daddy?' Because I think everybody understands that when we are fouling the Earth like this, it has concrete implications not just for this generation, but for future generations."[10]

Obama quickly declared that the Deepwater Horizon was not simply an isolated incident but rather a symptom of a bigger problem: the damaging consequences of a corrupt energy industry and the lack of regulatory oversight. By June 2010, in an Oval Office address, Obama accused federal regulators who handed out oil drilling permits of being controlled by big oil companies.[11]

Energy companies complained of a massive regulatory onslaught. By November 2010, the Environmental Protection Agency was said to be "developing and finalizing nearly 30 major regulations and more than 170 major policy rules." The American Legislative Exchange Council also noted that the volume of activity "had already surpassed the Agency's regulatory output in the entire first term of Bill Clinton."[12]

In December 2010, Obama ordered new drilling restrictions, placing "the entire Pacific Coast, the entire Atlantic Coast, the Eastern Gulf of Mexico, and much of Alaska off-limits to future energy production."[13] By canceling the 2010–2015 lease plan that allowed for new development on the Outer Continental Shelf, Obama caused a nearly three-year moratorium on new offshore drilling. He planned to push the ban further. "The Administration's draft five-year plan prohibits new offshore drilling and only allows lease sales to occur in areas that are already open."[14]

In January 2011's budget, Obama pledged to end oil and gas industry tax breaks, called "subsidies" by critics. These tax breaks weren't subsidies at all, according to American Petroleum Institute's Jack Gerard. "The federal government by no stretch of the imagination subsidizes the oil industry. The oil industry subsidizes the federal government at a rate of $95 million a day."[15]

If the oil industry felt under assault, the coal industry believed its very existence was at stake. "Over the past year and a half, we have been fighting President Obama's administration's attempts to destroy our coal industry and way of life in West Virginia," declared Governor Joe Manchin, a Democrat, in 2010. "We are asking the court to reverse EPA's actions before West Virginia's economy and our mining community face further hardship."[16]

Indeed, the regulatory push against coal and fossil fuels would cost thousands of people their jobs.

How closely tied was Barack Obama, his words and actions, to the health of the coal industry? The day after his reelection, November 7, 2012, shares in some coal company stocks slumped by as much as 10 percent.[17]

His cap and trade proposal, which would essentially tax companies for their carbon output, was controversial because of the costs it would impose on ordinary Americans. When a Republican Congress failed to pass cap and trade, Obama vowed to continue by other means. "Cap-and-trade was just one way of skinning the cat; it was not the only way," he proclaimed at a press conference just after the midterm 2010 elections when Democrats lost control of the House.[18]

Valuations of energy companies are influenced by many things including the price of oil, international demand, and world events. It is simplistic to say that President Obama, or any U.S. president, has the power to entirely dictate the health of an energy company. Many factors determine the price of energy, supply, and demand. And of course, also keep in mind that "there is a multi-year lag time between policy decisions and/or price signals, and subsequent changes in production."[19] In other words, some of his policy decisions' detriment to the fortunes of energy companies may be indirect and delayed.

That said, an American president can influence the perceived economic health of traditional energy companies. During his tenure in office, even to the very end, the Obama administration aggressively pushed for regulations and restrictions on the energy industry including coal companies, offshore oil drilling firms, and oil companies.[20] As Laurence Tribe of Harvard Law School pointed out, Obama's policies represented something never seen before by the coal industry. "Coal has been a bedrock component of our economy and energy policy for decades," he declared.

By "manifestly proceed[ing] on the opposite premise," Obama's energy policy "represents a dramatic change in directions from previous Democratic and Republican Administrations."[21]

Obama's policy statements and regulatory actions had dramatic effects on the valuations of traditional energy companies because his words and actions had major implications to their future cost of doing business. In some cases, valuations were driven down; in others, investors simply became cautious about what the future might hold. During this time, certain Obama friends profited by aggressively buying stock in those very same companies.

Valuations matter. Consider the fortune of coal companies. Set aside the questions about global warming and the merits of the coal industry and look at what happened. Between January 2009 when Obama took office and early 2015, shares of many coal companies plunged more than 90 percent. Several companies went bust. "Only the very toughest will survive," Sheila Hollis, the partner at the law firm Duane Morris who heads up energy, environment, and resources, told CNN.[22]

Yet, in the midst of that financial avalanche, many of Barack Obama's closest friends rushed in, buying up coal company shares. And while the exact price that they bought and sold their shares is not public information, many of them appear to have profited from well-timed investments. The same can to be said for other energy company stocks. When Obama imposed restrictions on offshore drilling in the United States, friends with Obama administration access bought shares of stock precisely in companies that work in the offshore oil drilling sector.

John Rogers, as we saw in the previous chapter, is one of those friends. Rogers became close to Barack Obama over the years, including as a financial supporter. Rogers has positioned himself as a "value investor." In other words, he looks for opportunities to buy stocks on the cheap and then sell them high. The key, of course, is finding stocks that are undervalued when you buy them.

By the third quarter of 2010, in the wake of Obama's regulatory push, Ariel Investments had bought stakes in Gulf Island Fabrication, which builds offshore oil platforms. The stock had been trading at more than $23 a share when Obama announced his ban on offshore drilling on April 30. Later that summer, Ariel bought it at $15. Rogers's firm also increased by 300 percent its shares in Mitcham Industries, a small Huntsville, Texas–based company that sells and leases seismic data equipment to the oil industry. Around the time President Obama had announced his ban on offshore oil drilling, a smaller Houston-based offshore oil-drilling company called Contango Oil and Gas saw its stock price drop from more than $55 a share to $43 (21 percent). Not long after, Rogers's Ariel jumped on more than 280,000 shares of the company's stock. As Obama continued to push new regulatory and tax rules on energy companies, Rogers's fund continued to buy up affected energy stocks on the cheap. By the end of 2012 Rogers's fund had bought close to a million shares in Team Industrial Services, an oil well maintenance and repair company, and shares in National Oilwell Varco.[23]

John Rogers's Ariel Investments bought its first large stake in Chesapeake Energy Corporation in the second quarter of 2011, gobbling up 1.6 million shares at more than $47 million. By the third quarter of 2013, Rogers's fund had purchased two million shares of U.S. Silica Holdings, which provides sand for fracking.[24]

Of course, any investor could have taken advantage of plummeting energy company stock prices, but any other investor would not have had the luxury of access to the policy makers who were driving down the prices. As Rogers's firm was making these large bets, he regularly visited the White House, enjoying the privilege of private meetings with those who were shaping policies targeting the energy industry. Between May 2011 and February 2014, Rogers visited the White House at least seven times for various meetings with Obama and energy-policy staff. [25]

On January 19, 2012, for example, he had a meeting in the Old Executive Office Building with Brian Deese, a senior adviser to his friend President Obama, who was responsible for regulations related to climate change and energy. Deese had served as deputy director of the National Economic Council with responsibility for energy policy early in the Obama administration. Deese and Rogers also served together on Obama's Financial Capability Advisory Council. [26]

Rogers also enjoyed access to Pete Rouse, counselor to the president. Rouse had been the former chief of staff to Barack Obama when he was in the Senate and was intimately involved in shaping Obama's energy policy. It was Rouse who "helped acquaint" Obama "with the nuances of energy policy," noted the *New York Times*, and played a major role in climate change and energy policy formulation in the White House.[27] Rogers and Rouse met in the White House on July 27, 2010, July 16, 2012, and then again on January 24, 2013.

Rogers met on December 21, 2011, with Gene Sperling in the West Wing. Sperling was the head of President Obama's National Economic Council and was also a key part of the president's "energy and environment team."[28]

There were also private meetings with President Obama himself—on April 11, 2013, in the Roosevelt Room, the meeting room at the heart of the West Wing, and on February 28, 2014, in the Oval Office. Rogers was also in frequent contact with Valerie Jarrett, President Obama's senior adviser and close friend. Rogers met with Jarrett in the West Wing on May 4, 2011, and again on January 23, 2013.

Rogers and Jarrett were so close that after she left the Obama White House, Jarrett joined the board of directors of Ariel Investments.[29]

Records of Rogers's e-mail, text, or telephone communications with Obama and his top aides were not available.

I contacted Rogers about these investment decisions and his meetings. Mellody Hobson, president of Ariel Investments, called me back and said that there was no connection between these meetings and any investment decisions that Ariel made. These were social visits, she said, with old friends. Business was not discussed.

It was not just longtime Obama family friends who were buying into energy company stocks. Some of his key financial backers were willing to make even bolder plays than Rogers's.

One of President Obama's largest financial supporters was hedge fund investor Tom Steyer. Born in 1957 in New York City where his father was a Wall Street lawyer, he enjoyed the perks of a wealthy upbringing. Educated at the elite Buckley School on New York's Upper East Side, he went on to Phillips Exeter and later Yale. Like his father he went into finance, working at financial behemoths Morgan Stanley and Goldman Sachs. While at Goldman he learned at the feet of eventual Clinton Treasury secretary Robert Rubin.[30]

Steyer eventually headed west, moving to San Francisco, and became immersed in the world of leveraged buyouts and private equity. In 1986 he founded the firm Farallon Capital Management. Over the next couple of decades, he built it into a large firm and accumulated a net worth north of $1 billion. He became increasingly involved with politics, raising vast sums for John Kerry in 2004 and Hillary Clinton in 2008.[31]

A major source of Steyer's wealth was financing coal projects in Indonesia and Australia. By injecting hundreds of millions of dollars into some of the region's biggest coal mines, investors stimulated production dramatically. Everyone made money, including Steyer. As employees at Farallon explained to Reuters, Steyer would have had to sign off on these coal deals. "The discretion to make or break any investment rested with him," remarked a Farallon investor, who requested anonymity.[32]

Steyer became an important and key financial backer for Obama's 2008 campaign. As *The New Republic* put it, "Hedge-fund billionaire Thomas Steyer threw one of the biggest Obama fundraisers of the entire campaign."[33] Funny to think that Obama was backed by coal money, but Steyer also became a true believer concerning Obama's environmental agenda. He increasingly spoke out on the dangers of global warming and the threats posed by the continuing use of fossil fuels.[34] This of course makes his stock trades in coal company stocks and related companies during Obama's tenure in the White House all the more interesting.

President Obama's war against the coal industry facilitated Tom Steyer and his fund's ability to buy coal-related stocks on the cheap. It could be seen as a neat and quiet payback. In early 2009 Farallon took a large stake in FreightCar America, a company that "specializes in the production of aluminum-bodied coal-carrying railcars." Farallon actually owned more than a million shares of stock, which amounted to almost 10 percent of the company.[35]

Steyer was only getting started. His Farallon Capital made a big plunge into Massey Energy in early 2011, scooping up more than 1.1 million shares.[36] It was Massey that was operating the Upper Big Branch Mine in West Virginia when twenty-nine miners were tragically killed in April 2010.[37] In addition to the fallout of this catastrophe, the company also faced industry consolidation, caused by the changing energy market and increasing regulatory pressures brought by the Obama administration.

The press against coal companies led to shutdowns and consolidations. By December 2011, *Fortune* magazine was reporting, "The list keeps growing. Closed and scuttled coal plants."[38] And that meant money could be made by buying them on the cheap. Farallon did well with its bet on coal. The price jumped from $53.65 a share on December 31, 2010, to $68.36 on March 31, 2011, as the company was in the process of being sold off to Alpha Natural Resources.[39]

Ironically, as Steyer made these trades and others, he was increasingly outspoken in his environmental views about global warming and the need to move away from investments in fossil fuels. In September 2012, Democrats gathered for a national convention in Charlotte, North Carolina, to prepare for the presidential election months away. Tom Steyer took to the podium and offered a ringing endorsement of Barack Obama for reelection. "President Obama knows that advanced energy is America's future," Steyer told Democratic National Committee delegates in Charlotte. "And my bet, as a business man, is that he's exactly right."[40]

He also admonished investors to stay away from fossil fuel investments. "During the last several years we've seen tremendous progress on new technologies that can make us energy independent and create thousands of jobs," he said. "This is about investing for the long haul, not for a quick and dirty buck."[41]

Not that Steyer was opposed to the "quick and dirty buck" for himself.

In January 2012, Steyer penned an op-ed for the *Wall Street*

Journal with John Podesta, arguing that the United States needed to restrict the importation of foreign oil because clean energy would fill the gap. He argued that the United States did not need to "build a pipeline to import more foreign oil" but should instead focus on alternative energy.[42] And indeed, President Obama decided in January 2012 not to approve the Keystone XL pipeline from Canada.[43]

Steyer argued that Keystone needed to be opposed because "the pipeline would completely change the rate at which the oil comes out of the ground."[44] But, of course, Steyer was heavily invested in another *competing* pipeline company that was already taking Canadian tar sands oil out of the ground.

Steyer's portfolio tells an interesting story.

Farallon's largest holding in the first quarter of 2012 was El Paso Natural Gas Company, a pipeline owner that was in the midst of being taken over by another pipeline operator, Kinder Morgan. Farallon owned a whopping twenty-two million shares in the company; it was by far their largest holding in the fund.[45] It is interesting to note that Kinder Morgan had its own pipeline connecting the Canadian tar sands to a port in the Pacific.[46] The Keystone XL pipeline was a potential rival as Kinder Morgan proposed an extended pipeline to compete.[47]

As Steyer was making these trades, he was increasingly outspoken about the moral necessities of dealing with climate change. He argued that climate change "is the issue we'll get measured by as a country and a generation. If we blow this, it will be because we were very focused on the short term, on our pocketbooks, and we had no broader sense of what we were trying to do and what we were trying to pass on."[48]

Steyer didn't stop with coal or pipeline companies. Farallon also grabbed shares of a number of other fossil fuel companies at this crucial time, including 100,000 shares of the oil drilling company Schlumberger, Ltd., 236,000 shares in Ultra Petroleum, and a large chunk of oil and natural gas producer Encana Corporation.[49] As these trades were going down, Steyer was the co-manager of Farallon.[50] He stepped down at the end of 2012.[51] After leaving Farallon, the firm still invested his money and he remained as the equivalent of a limited partner in the firm.[52]

Like Rogers, Steyer enjoyed high-level private access at the White House while his firm was making these trades. Between 2009 and 2012 he had multiple intimate meetings with the top leadership at the White House. According to White House logs, he met with Rahm Emanuel, then White House chief of staff, for six hours on the evening of September 29, 2009, and another meeting on March 4, 2010. In 2011 he met twice with the new White House chief of staff, Bill Daley. In 2012 he met regularly with Pete Rouse. In 2014 he had four meetings with White House counselor John Podesta.

There is of course no way of knowing what they exactly discussed and what other correspondence might have taken place. I contacted Steyer's office to ask about his stock picks and his meetings. His office vigorously denied that there was any connection between the two: "Any assertion that Tom Steyer benefited financially from material non-public information received from officials in the Obama administration is false."

George Soros first crossed paths with Barack Obama when the young state senator from Illinois decided to mount a campaign for the U.S. Senate in 2003. One of his supporters was the New York–based billionaire.[53] In December 2006, the two men met in New York for private talks. The following month, Obama announced his run for president. In a surprise move, Soros immediately announced his support, no doubt much to the chagrin of Bill and Hillary Clinton and their allies.

One of the most controversial financiers of the past century, Soros, born and raised in Hungary, lived under the Nazi occupation. He rather bizarrely proclaimed to *60 Minutes* that he felt no remorse for pretending to be the teenage, Aryan godson of a government official working with the Nazis to confiscate the property of Jews, saying, "I could be on the other side, or I could be the one from whom the thing is being taken away."[54] After the war, Soros moved to Great Britain and eventually worked his way to a position in arbitrage in a merchant bank in London. In 1956 he moved to New York City to work as a European stock analyst. By 1966, with the help of his employer, Arnhold and S. Bleichroeder, he established his first investment fund. He continued to build on his investment theory that markets could be leveraged based on emotions and political turmoil. In 1969, Soros had parlayed his successes enough to be entrusted with his own hedge fund, eventually building one of the largest and most successful investment hedge funds in the world, the Quantum Fund. Reinvesting his earnings, he grew to be one of the wealthiest investors on the planet.

Along with his new status as a billionaire, Soros became increasingly engaged in politics, identifying with the political left.

In the past, Soros did well financially and profited from political and economic chaos. In 1992, Soros correctly bet against the British pound and famously made $1 billion in a single day. Former U.S. Chamber of Commerce chief economist Richard Rahn reported the allegation that Soros received insider information obtained from the French central bank and the German Bundesbank when making that trade.[55] Soros was charged in 2002 by

French authorities for engaging in insider trading when he pocketed more than $3 million in profits from trading shares in the French Bank Société Générale.[56]

Soros and Obama shared common views on global warming, climate change, and the fossil fuel industry. Like Tom Steyer, Soros aggressively financed anti–fossil fuel causes. He was, in other words, a powerful ally in the "smash" campaign.

In 2009, he pledged to spend a cool $1.1 billion to fund climate change initiatives.[57] Money went to groups like Friends of the Earth, Alliance for Climate Protection, Earthjustice, the Earth Island Institute, Green For All, and the Natural Resources Defense Council, who were in the business of ringing the alarm bells about global warming and the need to go after the fossil fuel industry. The Center for American Progress, another major recipient of Soros's contributions, pushed aggressively for "green energy" to take down "King Coal."[58]

Soros continued trading aggressively in energy stocks. He bought into alternative energy companies, many of which stood to directly benefit from the Obama administration's stimulus program that was designed to help the alternative energy companies succeed.[59] But he also curiously loaded up on fossil fuel stocks, particularly after they had been "smashed" by Obama administration policies and groups that he himself was funding.

In the first quarter of 2012, Soros's Quantum Fund was loaded up with petroleum stocks. He held shares in Transocean Limited, an offshore oil drilling company, which had seen its share prices drop throughout 2011. Soros was able to grab them cheap. CVR Energy, an oil refinery, was one of his largest holdings. CVR had seen its share prices tank in late 2011 from $27 a share in October to $17 a month later. Soros's investment fund also held large shares in Marathon Petroleum, oil refiner Tesoro, Chevron, and Interoil.[60]

Early on in the Obama administration, one of Soros's biggest investments was in the Brazilian energy giant Petrobras, the stated-owned oil company. Because of President Obama's ban on U.S. offshore oil drilling, the Obama administration's Export-Import Bank approved large loan guarantees to Petrobras to explore drilling off the coast of Brazil.[61] As *Investor's Business Daily* noted at the time, "While the administration has imposed an almost total ban on U.S. offshore drilling, the U.S. Export-Import Bank has guaranteed billions in loans to Petrobras as President Obama has encouraged its offshore drilling efforts, promising to be Brazil's best customer for oil found with the help of our money."[62]

Soros considered coal companies, ostensibly, to be a major source of global warming, but he still bought up shares at the right price. Soros himself contributed aggressively to the smashing of coal company stock prices through his think tanks and political contributions, but once they were smashed, he kept up buying shares on the cheap. In early 2011, Soros held stock in International Coal Group.[63] In the third quarter of 2012, he bought more than four million shares in Peabody Energy.[64] By the summer of 2015, he bought one million more shares of Peabody Energy and half a million shares of Arch Coal. Six years earlier, those Peabody shares would have cost him about $90 each, but lo and behold, "under the Obama administration, which has punished the coal industry with costly mandates and regulation, Peabody shares have fallen to around $1."[65]

It is not possible to fully know how much Soros was trading in coal stocks after the smash had occurred, but as one financial analyst speculated to *Forbes*, "I think George Soros used the government like a blunt object to beat down coal stocks and make money shorting them."[66]

Speculation about Soros's involvement in selling coal stocks short came after Soros had bought into the companies, which, when made public, led to "a pop in stock price for both companies" (Arch Coal and Peabody Energy).[67]

Soros's firm also held $234 million in shares in the coal producing company Consol Energy, but slowly sold off those shares.[68]

Soros enjoyed the advice and input of his brother Paul, who thoroughly knew the fossil fuel industry, particularly the global market for coal and oil. Paul Soros was an innovative port-builder sometimes referred to as "the invisible Soros."[69]

In June 2014, Obama announced the proposed Environmental Protection Agency edict for a set of new guidelines that would essentially force many coal-fired power plants off-line before their end of life. This resulted in the shutting down of hundreds of coal-fired plants.[70] States like West Virginia and Kentucky were particularly hard hit. Some estimates showed fifty thousand lost jobs over a five-year period.[71] No doubt, many of these were due to the new regulations.

At the same time that the U.S. coal industry seemed to be going into death throes, Soros began buying up coal stocks cheap.[72] In just a few years, the Dow Jones U.S. Coal Index plummeted from $504 a share to less than $11.[73]

Soros and executives from his fund had regular meetings with senior officials in the White House who crafted administration regulatory efforts on energy policy. These included numerous meetings with Vice President Joe Biden, White House economic advisor Larry Summers, John Podesta, and top energy advisor Michelle Patron. There were also, of course, private meetings with President Obama, too.

Vice President Joe Biden and Secretary of State John Kerry sit with President Barack Obama as they meet with Chinese officials in 2015. Biden's son Hunter and Kerry's stepson Christopher Heinz had quietly struck multibillion-dollar deals over previous months with Chinese government-connected companies.

(POOL/GETTY IMAGES NEWS/GETTY IMAGES)

TOP: December 2013: The vice president arrives in Beijing with his granddaughter Finnegan and son Hunter aboard Air Force Two. Ten days after Biden left, Hunter's firm would seal a highly unusual $1.5 billion deal with funding from the Chinese government.

(POOL/GETTY IMAGES NEWS/GETTY IMAGES)

BOTTOM: Hunter Biden (left) in Beijing with his father. "Family comes first" Joe taught Hunter. Months after his father was sworn in as vice president, Hunter and Chris Heinz set up a private business and quietly secured billion-dollar deals around the globe with oligarchs and those closely connected to foreign governments negotiating with their fathers.

(POOL/GETTY IMAGES NEWS/GETTY IMAGES)

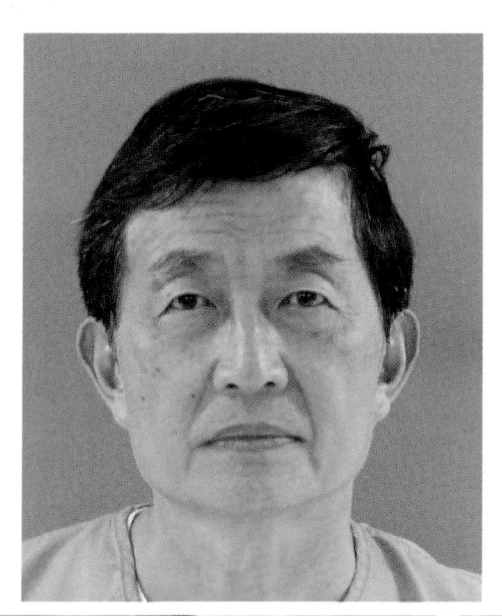

TOP: Engineer Szuhsiung "Allen" Ho pleaded guilty to sharing nuclear secrets with the Chinese government. The FBI charged Ho and his company, China General Nuclear Power Company (CGN), with stealing secrets with military application. CGN was owned in part by Hunter Biden and Chris Heinz's investment fund.

(KNOXVILLE SHERIFF)

BOTTOM: Ukrainian oligarch Ihor Kolomoisky. His company Burisma has made the Biden family a lot of money. The United States backed a $1.8 billion loan to the Ukraine, much of which flowed through Kolomoisky's PrivatBank. More than $1 billion mysteriously disappeared.

(VALENTYN OGIRENKO/REUTERS PICTURES)

TOP: Senator Mitch McConnell and Elaine Chao in 2011. The consummate Washington power couple, they have benefited financially from her family's extraordinarily close links to the Chinese government.

(MEDIAPUNCH INC/ALAMY STOCK PHOTO)

BOTTOM: Elaine Chao meets as a private citizen with Chinese vice-premier Liu Yandong. Her father and sister have served on the board of a company at the heart of the Chinese military-industrial complex.

(XINHUA/ALAMY STOCK PHOTO)

Pictured with then–Speaker of the House (and later disgraced) Dennis Hastert (center), former congressman Bud Shuster (left) passes the baton to his son Congressman Bill Shuster (right). Bud would go on to lobby his son's congressional committee along with his other son, Robert Shuster.
(DOUGLAS GRAHAM/CQ-ROLL CALL GROUP/GETTY IMAGES)

TOP: President Barack Obama and "first friend" Marty Nesbitt take in a basketball game together. Nesbitt generally kept a low profile during the Obama years but profited from his friend's regulatory actions in a series of "smash and grab" investments.

(MCT/TRIBUNE NEWS SERVICE/GETTY IMAGES)

BOTTOM: Marty Nesbitt playing golf with President Barack Obama. Nesbitt formed an investment firm that would follow in the wake of President Obama's regulatory actions. When those actions nearly destroyed companies Nesbitt would buy them after their shares plummeted. Today Nesbitt serves as the chairman of the Obama Foundation.

(POOL/GETTY IMAGES NEWS/GETTY IMAGES)

George Soros, billionaire investor and backer of President Obama. As the Obama administration "smashed" energy companies, Soros "grabbed" shares on the cheap. Other Obama insiders did the same.

(BRYAN BEDDER/GETTY IMAGES ENTERTAINMENT/GETTY IMAGES)

ABOVE: Jared Kushner (right) posing with his father, Charles. While Jared served as a senior White House adviser helping to steer Middle East policy, his father quietly tried to negotiate a large investment deal with the government of the United Arab Emirates.

(PATRICK MCMULLAN/GETTY IMAGES)

BELOW: Chinese president Xi Jinping meets with President Donald Trump in Florida. Jared Kushner and Ivanka Trump (left) have major business ties with politically connected companies in China.

(JIM WATSON/AFP/GETTY IMAGES)

Another one of Barack Obama's closest financial supporters was David Shaw, a computer scientist, computational biochemist, and advocate for alternative energy who launched a hedge fund in 1988.[74] Shaw's hedge fund was also well positioned in the wake of Obama's regulatory actions against oil companies. Shaw had been an early financial backer of Obama, serving as a campaign finance bundler who raised between $200,000 and $500,000.[75] He had donated more than $1 million to Obama's Organizing for Action committee and in 2015 gave between $250,001 and $500,000 to the Obama Foundation.[76]

Shortly after he entered the White House, Obama appointed Shaw to sit on the President's Council of Advisors on Science and Technology (PCAST).[77] The council issued a report in 2010 arguing for the United States to transform the country's "carbon-based economy into a safer, more sustainable, and economically advantageous energy ecosystem."[78] According to White House visitor logs, Shaw enjoyed direct access to the president, visiting him in the White House residence and the Oval Office for private meetings.

Not only is Shaw a two-time Obama bundler, but also he and his hedge fund employees have been significant donors to the Democratic Party.[79] Shaw just happens to be the largest shareholder of First Wind Holdings, LLC, the project sponsor of Kahuku Wind Power project, which snagged $117 million from Obama's DOE "junk loan portfolio."[80] First Wind, which received more stimulus funds, has another connection close to Obama: Larry Summers, head of President Obama's National Economic Council in the White House. Before he took that job, Summers was employed by D. E. Shaw and was paid $5.2 million over two years. He reportedly worked there just one day a week.[81]

His investment funds also bought heavily in the coal and oil sector. In late 2010, as coal stock tanked, Shaw's fund bought and held stock in more than ten coal companies, including Arch Coal, Peabody, Westmoreland Coal, and Patriot Coal Corporation. The fund also owned shares in more than twenty-five oil companies.

Shaw's fund was also aggressively moving on coal and oil company stocks as the Obama regulatory push was occurring. In 2011, as Obama's war on coal was picking up steam, the fund likely shorted major coal producers Alpha Natural Resources, Arch Coal, Consol Energy, Peabody Energy, Teck Resources, and Walter Energy.

By August 2015, D. E. Shaw and George Soros's Soros Fund Management, two hedge funds with close Obama ties, were ironically among the largest shareholders in Arch Coal.[82]

As the Obama administration was aggressively pushing more regulations and higher taxes on the fossil fuel industry, it was championing not only alternative energy but also natural gas. Obama declared on June 25, 2013, that natural gas is the future and argued that it was "the transition fuel that can power our economy with less carbon pollution."[83]

This, of course, created commercial opportunities for those connected to the Obama team. Cheniere Energy would become the first and largest U.S. exporter of liquefied natural gas (LNG).[84] Cheniere Energy is the owner of the Sabine Pass LNG export facility, a thousand-acre facility that straddles the Texas–Louisiana border. Sabine Pass became the first terminal to receive a final approval from the U.S. Federal Energy Regulatory Commission (FERC).[85] Getting approval, of course, required political connections. Cheniere had them.

Cheniere had a series of meetings at the White House in January 2013, just before they received Obama administration approval. The first meeting took place on January 14, 2013, the second on January 29, 2013, according to White House visitor logs. Cheniere CEO Charif Souki arrived in the White House along with company executives Patricia Outtrim, vice president of governmental and regulatory affairs, and Ankit Desai, vice president of government relations. The senior White House official who organized those meetings was Heather Zichal, who was serving as the deputy assistant to the president for energy and climate change. Zichal had worked with Desai, the Cheniere lobbyist before, on John Kerry's 2004 presidential campaign. Desai also formerly served as political director for then U.S. senator and now vice president Joe Biden. Having paid for their government relations, Cheniere got the first ever regulatory approval for an LNG export facility. Just after those White House meetings, Zichal resigned her White House job and made plans to "move to a non-government job."[86] She joined the board of directors at Cheniere.[87]

11

A REAL ESTATE MOGUL GOES TO WASHINGTON

- President Trump is not the first person at the highest levels of government to own great quantities of real estate with the risk of massive conflicts of interest.
- Penny Pritzker served as Obama's secretary of commerce; her companies leased property to the federal government, built up a Washington, D.C., real estate empire during her tenure, and did regular business with foreign governments.

President Donald Trump is a real estate mogul with myriad properties, companies, and branding deals, who has children with more of the same. During and after his successful campaign and into his presidency, many have expressed alarm over his forgotten financial disclosures, his foreign income sources, his unwillingness to divest, and his practice of doing business with many entities that would have reasons to curry favor with the U.S. government. Some said it was an unprecedented situation to have a president with a wide array of global assets that make conflicts of interest inevitable.

Trump's situation is not as unique to the executive branch as one might think from the current media. During the Obama administration, another real estate mogul with extensive holdings,

Penny Pritzker, held power in the executive branch as commerce secretary. Pritzker's business interests blurred in highly questionable ways with her role as public servant. Her story provides a blueprint for the type of transactions that Trump and his family should seek to avoid, and for which they should be monitored.

As a study in ethics, if not personality, introspective and reserved Penny Pritzker is a Trump prototype. The accomplished and hardworking heir to the Hyatt Hotels fortune, who has massive real estate holdings and a family engaged in a wide array of international businesses, Pritzker has already faced the sorts of ethics issues that Trump faces or may face. How she and the Obama administration handled her conflicts of interest offers a detailed model of failure from which the Trump administration could learn.

Pritzker's and the Trumps' lives have intersected in the past. In addition to being rivals and competitors, they were once partners. It did not end well.

Donald Trump's first major real estate deal in New York was with the Pritzkers.

> He bought an option to buy the old Commodore Hotel in 1977 for $1 for his first project in Manhattan. With the help of extensive tax abatements, he and his new partner, the Pritzkers, spent $100 million converting the dowdy Commodore into the glitzy Grand Hyatt, a development that signaled the revival of New York after the financial crises of the 1970's.
>
> But the two partners never really got along. Two arbitration proceedings failed to stem the bickering, and in 1993 Mr. Trump sued Jay Pritzker and the Hyatt Corporation. The company countersued seven months later. The two

sides managed to settle their dispute in 1995, with the Pritzkers agreeing to pay Mr. Trump's legal fees and to finance a $25 million renovation of the hotel.[1]

The Pritzker family has been compared to the Rockefellers and deemed "America's Rothschilds" by *Forbes*.[2]

But they are far less known than either of those families to the American public. In part that is because the clan has been notoriously secretive and shy about media coverage. The family dictum was declared by the late A. N. Pritzker, Penny's grandfather. "We don't believe in public business," he bluntly said.[3]

Penny Pritzker grew up in a family that over the course of the twentieth century accumulated a vast empire—largely based out of Chicago—including hotels, commercial real estate, and industrial companies. While the family's business was built on hard work and financial savvy, mafia connections helped, too. As Gus Russo records in his book *Supermob* about the Chicago mafia, "What is most relevant to the Pritzker role in the Supermob is the large number of transactions that involved known crime figures," including the "Capone syndicate."[4] Some might argue that the only way one could do business in Chicago at the time was by doing deals with the mob.

Penny grew up in California. Her father, Donald, managed the Hyatt Hotel chain early on but died when she was young.[5] That brought her into the larger sphere of family members.

"My father passed away when I was 13," she recounted, "and I didn't have a sponsor in the family. Eventually my uncle Jay took me under his wing, but it was my cousin Nick who really welcomed me into the family enterprise, made it okay for me to be there. The first thing you need if you want to work is a job, and Nick gave me a job."[6]

After college, Penny landed in the family's home base of Chicago and worked her way through the Hyatt empire. She would eventually go on to serve on the Hyatt corporate board, as well as other companies like TransUnion and Wm. Wrigley Jr. Company.[7] Her family also partially owned a bank that became entangled in the subprime mortgage scandal.[8]

Through Marty Nesbitt, she became acquainted with a recent Harvard Law School graduate named Barack Obama. At the time he was a lecturer at the University of Chicago Law School and an up-and-comer in local politics. Soon, Obama, his wife, Michelle, and his in-laws (Michelle's brother, Craig Robinson, and family) were regular guests at Penny's lavish Lake Michigan vacation house.[9] The Obamas, Nesbitts, and other friends also started taking annual vacations to Hawaii, where they stayed at the Pritzkers' Hyatt Regency on Waikiki.[10]

Over the next two decades, as Barack Obama climbed the rungs of political power—Illinois General Assembly, United States Senate, and the presidency—Penny Pritzker would play a central role in bankrolling his campaigns. Beyond Obama, she has been a generous giver to all sorts of candidates. By one count she has given to more than 119 federal-level candidates since 1990—mostly Democrats.[11] Specifically for Obama, she held numerous fund-raisers in "her modernist home and sculpture garden." She raised money for his 2004 Senate bid, and the victory celebration was held at the Hyatt Hotel in downtown Chicago. During the 2008 presidential campaign, she served as the finance chair and raised millions for his White House bid. In 2012 she was a bundler for his reelection bid, collecting more than $500,000 in campaign donations.[12]

Her fund-raising prowess led NBC's Chicago affiliate to dub her "Obama's sugar mama."[13]

In short, Pritzker helped make Obama's rise to power possible. *Fortune* magazine concurred: "there are those who would argue that he wouldn't even be in the Oval Office without the millions she raised to fight Clinton."[14] The *New York Times* was even more blunt: "Without Penny Pritzker, it is unlikely that Barack Obama ever would have been elected to the United States Senate or the presidency."[15]

Obama's victory in 2008 led many to expect that Penny Pritzker would play a role in the new administration. Commerce secretary was the obvious choice. And why not? Penny Pritzker was a savvy businesswoman, loyal, and smart, but an appointment in the new administration never came. Why Obama ultimately passed on Pritzker is the subject of dispute. Some say she had too many conflicts of interest and baggage, which would result in a nasty confirmation fight. Recall that in 2008 the world was in the midst of the financial crisis. There was widespread anger at Wall Street and the financial class in general. These sentiments were doubtless the seed to the Occupy Wall Street movement a few years later.

For Penny Pritzker, many of the issues raised by the financial crisis hit home. The Pritzker fortune had been erected thanks to very complex, but legal, tax avoidance techniques. As *Forbes* magazine put it, the Pritzkers operated "one of the grandest and most successful family tax-avoidance schemes ever." One tax expert explained it, "Some of their tricks were old-school—buying profitable but run-down companies rich in tax-loss carry-forwards, for example. But others were revolutionary. They housed their wealth in trusts, many of them in the Bahamas." They named their children and nieces and nephews as beneficiaries. "In an additional twist, the offshore trusts would borrow money for the purchases from a bank-type entity called ICA, itself owned by the Pritzkers' trusts."[16]

Another confirmation hurdle was the matter of the war raging between Hyatt Hotels and organized labor. Unions had been fighting with Hyatt for years, alleging that poor working conditions and the squelching of worker rights were common at the hotel chain. Labor unions had been a huge constituency for Barack Obama.[17]

Pritzker herself says she took herself out of the running because the Pritzker family had gone through a nasty feud, in which several members of the clan had sued Penny and other members alleging mismanagement of the family fortune. Penny and the others denied it. But the lawsuit led to the reorganization of the family empire. In late 2008 she was "legally obligated to the family to unwind Hyatt" and sell off assets.[18] One official said, "She fears problems with her confirmation based on past business dealings."[19]

Whatever the reason, Pritzker would not yet become commerce secretary. But she would remain a player in the Obama White House and continue to expand her real estate portfolio, which increasingly intersected with the federal government and the Obama administration.

President Obama appointed her to two economic advisory boards in the White House. The President's Economic Recovery Advisory Board was erected by Obama and included a handful of senior corporate titans from around the country and headed by two academics including a former head of the Federal Reserve. The board was designed to advise on economic matters.[20] Later, Obama created the President's Council on Jobs and Competitiveness and also appointed Pritzker.[21]

Throughout Obama's first term, Pritzker was a regular fixture at the White House, appearing at advisory board meetings and traveling there "for even minor White House events," in the words of the *New York Times*.[22] According to White House visitor logs, Pritzker visited the White House seventy-two times during Obama's first term.

At the same time, Pritzker launched a new real estate business in 2009 to be based out of the Washington, D.C., area. Artemis Real Estate set up shop on Wisconsin Avenue in nearby Chevy Chase, Maryland.[23] Describing itself as a "private equity real estate investment manager," the firm raised $736 million to be used to purchase office buildings. Much of the money came from government pension funds, including the Illinois Municipal Retirement Fund and the New York State Common Retirement Fund.[24] With her White House connections, the federal government, including the government department Pritzker would later head, became a profitable tenant for Artemis and the Pritzker empire.

Shortly after he was reelected in 2012, Obama offered Pritzker an appointment as his commerce secretary, which she readily accepted. Her financial disclosure, required by law, revealed the extensive scope and size of her financial empire. The 202-page disclosure listed holdings "in commercial and residential real estate, government bonds, art, casinos, timber, senior living communities, housing, an airplane leasing company, wood products, software and even agricultural land in Uruguay."[25] Later the document grew when she filed an amended disclosure of more than $80 million in income from trusts. It was a "clerical error," said her spokeswoman.[26]

The commerce secretary may not get the attention of the secretary of state or secretary of defense, but the office has enormous ability to help individual companies and industries. When the department started in the early twentieth century, its largest activity was managing America's lighthouses. But it quickly morphed into something much more powerful. The commerce secretary has a strong voice in the enforcement of product import and export restrictions, the selection of which tourism companies or areas get extra government attention, and the opportunity to help American companies access foreign markets. In general, they enjoy a large budget and vast regulatory powers. One scholar analyzed what the Department of Commerce spent its money on as Pritzker took the helm. "In Fiscal Year 2013, the Department of Commerce spent about $10 billion and employed 42,829 bureaucrats. A breakdown of the budget by function shows that some 30 percent goes to paying salaries, while 40 percent subsidizes private businesses and local development projects." The scholar's article is entitled "Department of Cronyism" for a reason.[27]

To deal with the looming problem of conflicts of interest, Pritzker boldly promised to step down from the board of the Hyatt Hotel corporation and other entities.[28] She also promised to sell off her interests in 221 companies. The moves were less dramatic than they sound. The 221 companies from which she divested were all relatively small. Her core holdings remained intact. She kept her $400 million worth of stock in Hyatt, and Hyatt remained a major government contractor.[29] She also kept the vast bulk of her real estate holdings, including the already advantageously positioned Artemis Real Estate. In her ethics letter to the assistant general counsel at the Department of Commerce, she did not commit to recusing herself from decisions affecting companies and entities in which she owned a stake, including Artemis. Instead, she offered a confusing and limited promise: "for a period of a year . . . I will not participate personally and substantially in any particular matter involving specific parties in which that entity is a party or represents a party." After a year, she would have the freedom to do whatever her vague promise covered anyway (so long as first authorized by federal regulation).[30]

As commerce secretary, she continued to appear to be directly engaged in the work of her investment company. At one point the *Chicago Tribune* reported that as commerce secretary she went to her investment company in River North, Chicago, where the staff hosted her as its "lunch and learn" guest.[31] This anecdote is but a tiny glimpse into a massive pattern of activity, largely unreported at the time in the mainstream media.

The steady intermingling among Pritzker's Department of Commerce, the Obama administration in general, and her real estate holdings showed, according to many, utter contempt for established ethical guidelines, both in magnitude and substance, even by Washington standards. As described in more detail below, her Department of Commerce would provide cash grants to companies that were tenants of her companies. Federal government agencies—including her own Department of Commerce—would pay rent to her companies.

Worse yet, her companies pushed aggressively into the politically sensitive markets of the Middle East and Russia as she served in the president's cabinet.[32] Hyatt Hotels (from which she had resigned her board position but not sold her stock) expanded overseas during her tenure, including new ventures to build three large hotels in Saudi Arabia.[33] Her D.C.-cozy company, Artemis (also from which she resigned her board position but was still an owner), would also serve as the landlord of politically sensitive foreign banks such as the Industrial and Commercial Bank of China, that country's largest state-owned financial institution.

Given the current bombardment of concerns faced by Donald Trump and his family, one would expect that similar concerns were raised before or during Pritzker's tenure. The public record is strangely silent.

At her Senate confirmation hearings, Pritzker was asked few questions about potential conflicts of interest or her family's financial dealings. "I was prepared that she was going to be attacked and prepared to help her," noted Senator Claire McCaskill of Missouri. "It was a lovefest." She was confirmed by the Senate 97–1.[34]

Penny Pritzker and her family understood the lucrative nature of leasing agreements with the federal government. In Chicago, the Pritzkers financed the building of a ten-story office complex on Chicago's West Side. In 2006, the FBI leased the building for a contract of $280 million in rent over the next fourteen years—more than twice what the building cost to be built. The Government Accountability Office (GAO), the investigative arm of Congress, singled the deal out as a particularly egregious example of government waste—costing taxpayers some $40 million.[35] Of course, one person's waste is another person's profit.

Secretary Pritzker promised bold leadership for the American business community. She promised to help make American companies more competitive overseas and to deal with the challenges and opportunities of technological innovation.[36] Meanwhile, her real estate companies began acquiring commercial properties that leased office space to the federal government.

In 2013 Artemis bought the Carlyle Center, a massive structure of brick and glass located in Alexandria, Virginia.[37] One of their tenants was the U.S. Patent and Trademark Office, which is under the U.S. Department of Commerce.[38] That meant that the commerce secretary could be seen as, in effect, the landlord of the Department of Commerce. The annual rent: $1.4 million.[39]

In July 2013, Artemis Real Estate bought a commercial building near Boston as part of a joint venture.[40] A U.S. government tenant was paying almost $670,000 annually in rent.[41] In October

2014, Artemis Real Estate partnered to purchase a property in San Rafael, California.[42] The Bay Area office complex had a U.S. government tenant who paid rent of $517,000 a year.[43] In September 2016, Artemis was partners in a venture to buy a building on the expansive Warren Parkway in Dallas.[44] A federal government tenant was paying $453,000 a year in rent.[45]

Artemis is potentially only one of several Pritzker real estate companies with federal government contracts. Pritzker's commercial real estate holdings go well beyond Artemis. But they are largely held in obscure limited liability companies that are difficult to trace. It is impossible to know the full extent of the federal government leases from which Pritzker collects money.

But it is not just about government leases. Pritzker's Artemis also acquired properties and did business with contractors to the Department of Commerce and with companies over which she had powerful regulatory authority. In October 2013, Artemis joined with a company called Onyx Equities to acquire corporate offices in Morristown, New Jersey.[46] The Mount Kemble Corporate Center was a massive 230,000-square-foot building. Two-and-a-half years later, the tech giant Avaya relocated its regional operations base into the building.[47] Avaya, which has an Avaya Federal Solutions division, was and is a Department of Commerce contractor. It also does work with other government agencies.[48]

Some tenants of Pritzker-owned companies also received Department of Commerce contracts and money from Pritzker. In August 2013, Artemis bought an industrial space in Huntington Beach, California. The 102,000-square-foot building didn't sit empty for long. It was quietly announced that a ten-year lease valued at $9.1 million was inked with Driessen Aircraft Interior Systems, Inc. Driessen designs, builds, and markets "high-quality galleys for commercial and private aircraft" and planned to occupy all of the space. Driessen is a subsidiary of Zodiac Aerospace, a company with more than twenty thousand employees. They had $3 billion in sales in 2012.[49]

Zodiac Aerospace is regulated by the U.S. Department of Commerce. "The Customer acknowledges that some of the Products may be subject to export laws and regulations such as laws and regulations issued by the U.S. Department of State International Traffic in Arms Regulations (ITAR), U.S. Department of Commerce Export Administration Regulations (EAR) or any other trade control regulations of any other country."[50]

Zodiac of North America, another subsidiary of Zodiac Aerospace, was awarded almost $800,000 in contracts by the Department of Commerce while Pritzker was secretary of commerce. Zodiac also did business with Commerce before Pritzker's appointment.[51]

On March 9, 2016, Zodiac Aerospace was picked by the Department of Commerce to participate in Obama's TechHire program.[52]

In March 2014, Pritzker's company Artemis took a stake in the construction of a $120 million office project on Capitol Hill in Washington, D.C. The construction of the large glass building on North Capitol Street just blocks from the U.S. Capitol Building was described by the *Washington Business Journal* as a "speculative office development" and a "risky venture." They had no preconstruction leasing commitments from tenants.[53] But when the project was nearing completion in 2015, it was announced that despite the "gamble" of building the project, they had two new tenants. They just happened to be the National League of Cities and the National Association of Counties.[54] Both lobbying organizations dealt extensively with the Department of Commerce, which handed out grants and cash to cities and municipalities around the country.[55] The National Association of Counties listed as a top legislative priority in 2015 the Department of Commerce grant program.[56] The new tenant at the Pritzker property was also involved in joint forums held by the National League of Cities and the Department of Commerce.[57]

Artemis went so far as to hire the federal government's top real estate executive to come work for them during her tenure.

In March 2015, Pritzker's Artemis hired the head of the General Services Administration (GSA) to come and work at the company. Dan Tangherlini left the GSA, where he managed real estate for the federal government, and took a job as the COO at Artemis Real Estate Partners. The *Washington Post* reported that Tangherlini has connections at nearly every level of government associated with real estate in Washington.[58] The administrator of the GSA oversees the procurement of real estate, travel services, and technology for the federal government.[59]

Then there was the question of money flow and favors involving Pritzker's family members. On March 30, 2015, the Chicago-based Clean Energy Trust, which supports clean energy start-ups through business development, won a piece of the $10 million in grants given by the U.S. Department of Commerce.[60] The trust was the only Chicago organization to receive federal money, and it planned to use the money for marketing, fund-raising, and staff.[61]

The cochair of the Clean Energy Trust is Penny Pritzker's cousin Nick Pritzker, of whom she spoke so fondly from her childhood.[62]

12

THE TRUMP PRINCELINGS

- Donald Trump has promised to drain the swamp.
- Among the challenges he may face are potential conflicts of interest and Princeling-style deals involving foreign entities and his children.

Donald Trump and his family arrived at the White House with countless business and personal relationships around the world—some going back decades. By one account, there were at least 111 Trump companies that had deals in at least eighteen different countries, including projects in Saudi Arabia, Indonesia, India, and Panama.[1] His contentious election to the White House has brought out apocalyptic predictions from his political opponents, and some concerns even from allies who would have him avoid the kind of ethical land mines that could weaken and distract his administration.

In an effort to steer clear of conflicts of interest during his presidency, President-elect Donald Trump had said that his hundreds of businesses would be placed in a trust managed by his adult sons.

As we have seen in previous chapters, foreign entities and even some U.S. corporations see the children and other family members of powerful American politicians as useful channels for monetary influence. As I said in December 2016, before Trump took office, "Foreign entities who are going to try to curry favor or get leverage over an American president often do it by trying to strike a sweetheart deal with a family member, often with one of the kids . . . That's my concern: you're going to have Bahrain, or Saudi Arabia, or Russia, or Kazakhstan, or some foreign power that wants something from Donald Trump is going to offer . . . a sweetheart deal to one of the kids. The kids will take it, and now, suddenly, there is some sense of either obligation or leverage or some form of embarrassment that can be used against the president."[2]

Or the Trump kids might not just be offered a deal; they might seek them out as the Biden and Kerry families did during the Obama administration. I also expressed concerns about Trump's continued involvement with his companies in a *Washington Post* op-ed piece I cowrote with Obama's "ethics czar" Norm Eisen.[3]

Both the Kushners and the Trumps operate in the world of real estate, which is highly political. Particularly overseas, foreign governments and oligarchs can make or break large real estate deals. From Russia to China to Kazakhstan, zoning, permits, and financing for a major hotel or golf course require the help and blessing of political figures.

So how might such a deal go down in a Trump administration?

As of this writing, the Trump administration has been in office for almost a year. Where do the possible conflicts and vulnerabilities lie? How do we hold the kids to account? Where might we have foreign or American entities employ the Princelings model to curry favor with the new administration? Studying the business ties of Trump and his children is a first step toward answering these questions.

We will begin with Jared Kushner and Ivanka Trump because

they have taken positions in the West Wing, influencing and shaping American policy.

JARED KUSHNER AND IVANKA TRUMP

Jared Kushner married into the Trump family from a real estate empire of his own. His father, Charles Kushner, and other family members developed their business primarily focused in New Jersey, including apartment buildings and commercial real estate. The year after Jared graduated from Harvard in 2003, Charles was arrested on charges of tax evasion, illegal campaign donations, and witness tampering. The U.S. Attorney leading the prosecution at the time was Chris Christie, who would later go on to be governor of New Jersey and serve in the Trump administration as the antidrug czar.[4] On March 4, 2005, Charles Kushner was convicted and sentenced to two years in federal prison. Four years later, in 2009, Jared's uncle Richard Stadtmauer, who was married to Charles's sister and who helped run the real estate empire, was sentenced to three years in federal prison for illegally writing off millions of dollars in charitable and political donations on company taxes.[5]

After his father went to prison, Jared assumed additional responsibilities for Kushner Companies. In January 2007, the same year that he met Ivanka, he made his largest bet, one that would come back to haunt him and would create a major source of financial vulnerability for the family. The real estate market was on the downturn after the massive boom that had dominated for almost a decade, when Jared made an ambitious move: using loans, he bought the 666 Fifth Avenue building in New York. The aluminum-jacketed office tower covers the whole block between Fifty-Second and Fifty-Third Streets in midtown Manhattan, near the famous Rockefeller Center.

Kushner Companies seriously overpaid for the property. The Kushners paid $1,200 a square foot, twice the previous record of $600. As the real estate publication *The Real Deal* put it, the transaction was "the priciest single building purchase in U.S. history."[6] The Kushners did not put a lot down on the deal, financing the purchase with a $1.2 billion loan from Barclays Capital and an additional $535 million in short-term debt. They were heavily leveraged.[7] Later in 2007, to offset the debt, Kushner Companies sold its entire portfolio of rental apartments for about $1.9 billion to AIG and Morgan Properties.[8]

The following year Jared Kushner took the helm as CEO of Kushner Companies.

When the stock market crashed in September 2008, the cash flow from 666 was not enough to cover the debt they still owed to the lenders. Jared had expected to make $120 million annually in rent. The property was bringing in $30 million a year with a 30 percent vacancy. To help cover the deficit, Jared sold the retail "condominium" portion of the property to the Carlyle Group and Crown Acquisitions for $525 million. But the move was simply buying time. By 2011 the Kushner Companies faced a default deadline on the property. The following year Vornado Realty Trust agreed to a $707 million deal for the retail portion of the property. Jared Kushner's company remains highly leveraged on the property. The interest-only mortgage is due in February 2019.[9]

What makes Kushner's situation precarious is not simply the financial vulnerability of his company but the fact that once his father-in-law was elected, Kushner sought and received a senior position in the White House. He has become one of the president's closest advisers. He has at various times described his role as a liaison to the business community while also positioning himself as someone involved in Middle East policy.[10]

When President Trump welcomed Chinese president Xi Jinping to his Palm Beach club Mar-a-Lago, Beijing used Jared Kushner as the back channel means of communication. "Since Kissinger, the Chinese have been infatuated with gaining and maintaining access to the White House," Evan S. Medeiros, an Obama administration senior director for Asia, told the *New York Times*. "Having access to the president's family and somebody they see as a princeling is even better."[11]

According to the *South China Morning Post*, Chinese diplomats see the "strategic use of the Kushner channel" as a good avenue to pursue better relations. As the paper puts it, "It is well known in Washington diplomatic circles that Cui Tiankai, the Chinese Ambassador to the U.S., has maintained close relations with Kushner and Ivanka Trump since Trump took office."[12]

The Chinese government has already tried to court the Trumps through the goodwill of small benefits. In March 2017, the Chinese government finally granted the family thirty-eight trademarks for various Trump projects after more than a decade of refusing to approve them. On the day that President Xi had dinner at Mar-a-Lago, China approved Ivanka's company's request to sell handbags, jewelry, and spa services in China.[13]

Ivanka Trump enjoys deep commercial ties with politically connected Chinese companies. Between 2013 and 2015, her company had clothes produced by a company owned by the Chinese government. Since her father was elected president, she has pledged to avoid business with state-owned companies. But tracking those ties is not always easy. Ivanka Trump shoes are also produced annually by a company called Huajian Group in China's Jiangxi Province. The company was founded in 1984 by a former PLA officer named Zhang Huarong, who today is deeply connected to the Chinese Communist Party power structure. He's a member of the Chinese People's Political Consultative Conference, an exclusive high-level advisory body for the party.[14]

These relationships hardly equate with the large private equity deals and real estate partnerships that Biden and Kerry family and allies struck with the Chinese, but we are still on the front end of the Trump administration. The Chinese and other governments will certainly be looking for ways to offer the Trumps and Kushners other favors and deals.

Even more problematically, the Kushner family appears to be *seeking* lucrative business deals given some financial icebergs that are on the horizon.

Jared Kushner needs an infusion of cash or he risks losing the 666 building. During his father-in-law's campaign and since he entered the White House, Kushner has sought deals with foreign entities to help him out with the property. We can expect that those from whom he is seeking money will seek policy favors in return.

Jared Kushner negotiated a tentative deal with China's Anbang Insurance Group, a financial institution in China headquartered in Beijing and "one of China's most politically connected companies," according to the *Financial Times*.[15] The chairman of the Anbang company, Wu Xiaohui, was married to the granddaughter of Deng Xiaoping, once the paramount leader of China.[16] Kushner's negotiations with Anbang took place both before and after Trump was elected president. Anbang withdrew from the negotiations when the deal was exposed by the *New York Times* and questions were raised about conflicts of interest.[17] A few months later, Wu Xiaohui was detained by Chinese authorities.[18]

Kushner and his father, Charles, have also sought a half-billion-dollar deal from a billionaire from the tiny Middle Eastern nation of Qatar. The nature of those discussions was highlighted by *The*

Intercept, an investigative publication with a very solid track record of journalism. According to a report by *The Intercept*'s Ben Walsh, Ryan Grim, and Clayton Swisher, Sheikh Hamad bin Jassim al-Thani, known as HBJ for short, held extensive negotiations with both Kushners. Throughout 2015 and 2016, while the campaign was going on, Jared and his father negotiated directly with HBJ to refinance the 666 property. Those negotiations continued through the spring of 2017—after Trump occupied the White House and Jared had joined him as a senior adviser.[19]

HBJ is the former prime minister of Qatar and ran the country's sovereign wealth fund. A former emir of Qatar once said of him, "I may run this country, but he owns it."

According to *The Intercept*, HBJ agreed to cough up a half billion dollars on the condition that the Kushners raise the additional funds needed to upgrade the property. Anbang was supposed to provide some of that funding before it pulled out.

Then in the spring of 2017, the Trump administration supported an effort by Qatar's regional rivals, including Saudi Arabia, to isolate the tiny nation. Arguing that the country was supporting terrorism—and there was plenty of material to support that argument—the Trump administration backed efforts by Saudi Arabia, the United Arab Emirates, Egypt, and Bahrain to block media outlets. It precipitated a diplomatic crisis in the region.[20]

Curiously, Kushner was fingered for backing the push against Qatar and reportedly delivered remarks written by the UAE ambassador to his father-in-law.[21]

During the presidential transition in December 2016, Kushner also held secret meetings with a Russian bank named Vnesheconombank. The state-owned financial institution is close to Vladimir Putin, has played a role in a past espionage case, and has been under U.S. sanctions since 2014. Kushner met with Sergey Gorkov, the bank's chief executive and a graduate of the Russian Academy of the Federal Security Service, or FSB. The FSB is the domestic intelligence arm of the Russian government. Kushner claimed that the meeting was generally a diplomatic matter. Bank officials said that the meeting was with Kushner in his capacity as the head of his family's real estate business.[22]

In short, Kushner has met with a veritable United Nations of lenders since his father-in-law was elected president. All are either government owned or deeply politically connected in their countries.

Another area of concern? Jared Kushner's involvement with Thrive Capital, an investment fund run by his brother, Josh. Unlike his older brother Jared, Josh Kushner appears to have little interest in politics. "Josh doesn't want a public profile," said one investor in Thrive, Darren Walker, president of the Ford Foundation. "He assured us that he would remain focused on Thrive."[23]

Jared Kushner has served on the Thrive board since 2009. On his personal financial disclosure, he listed capital gains from Thrive entities of over $5 million in 2016.[24] Thrive has invested in several businesses that could prove to be conflicts of interest. Among their holdings: OpenGov, a technology company that is providing tools for financial information on government spending.[25] Transparency is a good thing, but are there actions that Jared Kushner could take in the Trump administration to benefit the company? Thrive is also heavily invested in Oscar Health, a health insurance company. The company was founded by Josh Kushner. "Oscar is a health insurance company that employs technology, design, and data to humanize health care."[26] Oscar Health, of course, stands to win or lose based on decisions made about reforming Obamacare. How deeply involved is Jared Kushner in shaping American health care policy?

One of Trump's signature issues during the 2016 campaign was building a border wall with Mexico. An Israeli company with ties to both Kushner and the Trumps is hoping to get some of the substantial contracts for that massive construction project. Magal Security Systems is an Israeli security company that constructed a security fence in Gaza. Shortly after Trump was sworn into office, executives from the company arrived in the D.C. area for a security conference that involved the Department of Homeland Security and would no doubt include discussions on their barrier system. The day after Trump allowed that a security barrier such as theirs could all but stop illegal border crossings, shares in the company jumped 5.6 percent. The company's shares had already climbed nearly 50 percent since Trump's election in November.[27]

Who is behind Magal? The children of one of the company's former board members who, through various entities, is still a major stockholder at 44 percent, have purchased nearly $60 million in Trump condos.[28]

Kushner also has close financial ties to the Steinmetz family, an Israeli family that made their money in the international diamond trade. Beny Steinmetz is under investigation in the United States for allegedly bribing foreign officials in exchange for mining rights. Kushner and Beny's nephew, Raz Steinmetz, are partners on at least fifteen properties in Manhattan and the Trump Bay Street project in New Jersey.[29]

These are investments that Jared Kushner disclosed. But there are serious questions about how accurate those disclosures were. Since he filed his financial disclosure in March 2017, he has amended the document an astonishing thirty-nine times—adding seventy-seven investments he previously omitted.[30]

ERIC AND DONALD TRUMP JR.

Eric and Donald Trump Jr. chose not to join their father in the White House. Instead, their role has been to run the Trump-family empire. Still, it cannot be overstated that foreign governments and businesses are going to be eager to grant favors to the Trump Organization because they expect that it will help them with the president. Their expectations are certainly not grounded in America's best interest, and the appearance of such collusion is both demoralizing to American citizens and encouraging to corrupt players around the world.

Even before he was elected, we saw this happen when President-elect Trump spoke with Argentina's president. A previously stalled Trump building project in that country was revived when the government granted the necessary building permits.[31] The Argentine paper *La Nacion* reported that "Trump asked for them to authorize a building he's constructing in Buenos Aires, it wasn't just a geopolitical chat." Spokesmen for both presidents deny the subject came up.[32] And in many ways, it didn't need to. Foreign governments know that money talks. By helping American politicians and their families in the pocketbook, they are more likely to get favorable outcomes when they talk to them about other matters.

Just weeks after Donald Trump was inaugurated, a Chinese businesswoman named Angela Chen, who also goes by the names Xiao Yan Chen and Chen Yu, bought a penthouse in Trump Park Avenue, a condo on Manhattan's Upper East Side, for $15.8 million. Formerly occupied by Ivanka and Jared, Chen purchased the property from Trump Park Avenue, LLC, an entity controlled by the Trump Organization, so the deal effectively resulted in money in President Trump's pocket, or estate. Curiously, the penthouse was not on the market at the time and had no listing price.[33]

According to Jonathan Miller, a real estate appraiser in New

York, Chen overpaid for the condo. "In my view, the seller did a little better than what the market conditions would suggest."[34]

Besides the price, what made the deal a red flag was the buyer's ties with the Chinese government. Chen runs a firm called Global Alliance Associates, which describes itself as having developed "a well-deserved reputation as a respected and sought-after advisor on establishing business relationships in China."[35]

Chen also chairs something called the China Arts Foundation, started by Deng Rong, the daughter of former Chinese leader Deng Xiaoping.[36] Deng Rong served as the vice president of the China Association for International Friendly Contacts (CAIFC), an affiliate of the intelligence and foreign propaganda division of the People's Liberation Army.[37] CAIFC has been called a front for Chinese intelligence by academics and government experts. Its offices are reportedly located in a Beijing compound used by military units.[38]

Another pattern began shortly after Trump was elected. The governments of Bahrain and Azerbaijan held national and religious celebrations, respectively, at the Trump International Hotel, just blocks from the White House.[39] Trump has wisely promised that once he was elected any profits from such visits by foreign officials would be donated to federal coffers.[40] This is great but requires intense monitoring. There can be no doubt that these governments expect that by throwing business at the Trump family they will get favorable treatment in return. As we have seen throughout this book, in many foreign regimes, this is how political business is done. Trump, as U.S. president, and his children, as the first family, must maintain the diplomatic standard that this is not how political business is done in a constitutional republic.

Like Kushner, the Trumps have financial vulnerabilities. They owe $300 million in loans to Deutsche Bank, the German financial behemoth that has run into trouble with the U.S. Department of Justice over its banking practices.[41] The Trump Organization also owes hundreds of millions to the Bank of China for its real estate projects.[42]

For those keeping track, the Bank of China has already positioned itself well by partnering with the Bidens, placing the sister-in-law of the Senate majority leader on its board of directors, and now is a major lender to the president of the United States' company.[43]

Biden and McConnell aside, the question for Trump is: How will these loans play out? Will the Trumps be pressured or encouraged by their bankers to carry out policies for their benefit? It is unclear at this point and a game of vigilance. Some might argue that loans by themselves do not amount to much leverage. As John Maynard Keynes once said, "If you owe your bank a hundred pounds, you have a problem. But if you owe a million, it has." Still, if you have to refinance your loans, you might be offered favorable rates . . . with silk strings attached.

In January 2017, President Trump's lawyers offered assurances to the American public that the Trump Organization would not pursue new deals overseas. But the family business has continued to move forward with projects that it had previously negotiated. So the Trump commercial empire continues to expand, even while he occupies the Oval Office. Two projects in Indonesia will bear the Trump name. The project partners include Hary Tanoesoedibjo, who goes by Hary Tanoe, and who has deep political ties in Indonesia. Tanoe attended Trump's presidential inauguration as a guest of the Trump Organization. He recorded the event on Instagram with pictures of himself with Don Jr. and Eric.[44] Again, vigilance will be required by the first family to maintain goodwill without cultivating foreign expectations.

A new Trump office building and residential development in India is also under way. The Trump Organization's partners in the deal, IREO and M3M India, have sketchy histories. Both have been targets of anticorruption investigations by the Indian government. The investors behind IREO prefer anonymity: the funds for the project provided by IREO arrived via accounts in Mauritius and Cyprus, where their true identities are guarded secrets.[45]

The Trump financial empire is truly global, which means that President Trump is vulnerable to financial pressure by foreign entities. If he takes actions as president that are unfavorable to a country, his family estate could suffer. Consider what happened after Trump proposed banning people from entering the United States from seven predominantly Muslim countries, including Turkey. President Erdogan of Turkey called for Trump's name to be removed from the Trump Tower in Istanbul. Trump's name was not taken off the property. But had it been removed, Trump's licensing fee for that property would have been put at risk. According to financial disclosures, he earns fees of between $1 million and $5 million a year for that property.[46]

Ironically, for those who would criticize Trump for anything short of complete divestiture, the fact that he still owns the Trump Organization is exceedingly helpful for transparency. Remember that the adult children of politicians don't need to reveal their financial transactions to the public. So if the Trump Organization were owned by the sons and not the president, ethical watchdogs would have no way of tracking transactions involving the business. Sweetheart deals could be struck with zero accountability, as we have seen repeatedly in previous chapters. But because the president of the United States remains an owner, he will be required to reveal in his annual disclosure any transactions, debts, or assets that he holds.

As with Kushner, it is worth looking at the cluster of financial relationships that Trump's sons Eric and Donald Jr. have and their temptation to seek or accept lucrative deals from foreign or domestic entities looking to curry favor with the Trump administration.

In the rough-and-tumble world of New York City real estate, you can end up doing business with some very sketchy people. For decades, New York real estate has been a popular place for oligarchs, criminals, and shady companies to park their assets. As the *New York Times* revealed in a series of investigative reports beginning in 2015, foreigners with troubling histories have placed their money in expensive residential and commercial real estate as a means of hiding their assets. Alleged Russian mobsters and politicians, questionable Chinese businessmen, and likely corrupt foreign officials from Kazakhstan, Colombia, Malaysia, Mexico, and more have poured millions into New York real estate through secretive limited liability companies that make it hard to determine who actually owns the properties. Famous New York luxury buildings, like the Time Warner Center, have become havens for such people.[47]

Well before he decided to run for president, Donald Trump and his company conducted business affairs around the world and ended up doing business with numerous individuals with troubling pasts. Take a single project, Trump SoHo, a modern, stylish $450 million forty-six-story hotel condominium on Spring Street in New York. The Trump Organization partnered with a development company called the Bayrock Group, which provided capital for the project. The project was completed in 2010.[48]

The Bayrock Group was founded by Tevfik Arif, a Kazakh real estate mogul from the former Soviet Union. In 2010, Arif was arrested in Turkey and charged with running a prostitution ring and engaging in human trafficking. The charges were eventually dropped.[49] Arif had political ties in his home country, including business ties with Viktor Khrapunov, a former Kazakh energy minister and former mayor of the city of Almaty. Khrapunov was allegedly involved in a massive money-laundering scheme.[50] As the *Financial Times* reported, "Among the dozens of companies the Almaty lawyers say the Khrapunov laundering network used were three called Soho 3310, Soho 3322 and Soho 3203. Each was a limited liability company, meaning their ownership could easily be concealed."

The companies were created in April 2013 and used to buy apartments that corresponded with their names at Trump SoHo. The *Financial Times* added, "On the face of it, Mr. Trump was not a beneficiary of the apartment sales. The vendor was another limited liability company, Bayrock/Sapir Organization, LLC."[51]

Also involved in the Trump SoHo project was another Russian émigré named Felix Sater. Sater emigrated from the Soviet Union to Brooklyn when he was eight. According to Sater, he and his family fled the Communist country because of the persecution of Jews. Sater started his career as a stockbroker but then lost his trading license in 1991 after an incident in a hotel where he stabbed a commodities broker with a margarita glass. He spent a year behind bars. He left prison with a wife and child to support and hooked up with a childhood friend who was operating a Mafia-linked brokerage firm. Sater pleaded guilty in 1998 to one count of racketeering as part of a $40 million stock fraud scheme. They had been artificially inflating the prices of stock. Sater was connected to the Trump SoHo project in that he worked for Bayrock, the Trump partner in the property. Sater reportedly carried around a business card in 2007 in which he listed himself as a senior adviser to Donald Trump.[52]

It was the end of January 2017, and three men sat down together at the Loews Regency on Park Avenue: Felix Sater, an alleged Russian-American mobster, federal informant, and off-again-on-again Trump business partner; Michael Cohen, President Trump's personal attorney; and Andrii V. Artemenko, a dissident Ukrainian lawmaker. Artemenko had brought a proposed peace plan for his embattled nation, one he claimed enjoyed the blessing of Vladimir Putin via "top aides," and he sought to place it before the newly inaugurated president.[53] The meeting had the hallmarks of a back channel around Washington red tape, but Artemenko's choice of intermediaries is troubling. Two of the men had an interesting recent history.

On November 3, 2015, Sater e-mailed Cohen: "Our boy can become president of the USA and we can engineer it . . . I will get all of Putin's team to buy in on this, I will manage the process." Sater claimed, in the same communications, to have gotten Ivanka Trump access to Vladimir Putin's private office during a previous Moscow business trip. As reported by the *Washington Post* and the *New York Times*, the e-mail was in the context of a nascent and considerable Moscow real estate deal involving the Trump brand. Sater claimed he secured financing from VTB Bank, an entity sanctioned by the U.S. government. A letter of intent had been signed even as Trump's campaign for the highest office in the land gathered steam for the primary season. Permits and other favors were needed in Russia. Sater pitched a Trump visit to Moscow to "tout" the deal. In return, Putin would publicly praise Trump. According to Sater, the arrangement had a dual, and one might say "yuge," upside: Trump would be part of an historic real estate deal in the Russian capital. If that weren't enough, the deal would also clear Donald Trump's path to the White House. "If he says it we own this election." As of this writing, no one is saying why

Sater was so confident the two outcomes were connected. His price? He indicated his "home run" would be getting the U.S. ambassadorship to the Bahamas out of the deal.

Putin and Trump would find nice things to say about each other during the remainder of the campaign, but the deal went nowhere, petering out in January 2016.[54]

No one knows (at this writing) whether the deal provides any context or prologue for the now infamous June 2016 meeting between the Trump campaign and a Russian delegation promising *kompromat* on Hillary Clinton. Cohen would downplay Sater and his credibility to the *Post* and the *Times*, dismissing his talk of a direct pipeline to Putin as "salesmanship."[55] *New York Magazine* reporter Andrew Rice states Sater and Cohen knew each other as teenagers—perhaps explaining Sater's excitement, expressed to Cohen: "Can you believe two guys from Brooklyn are going to elect a president?"[56]

Yet that begs the question: In the face of the failed real estate deal, and his failed bravado, why did Cohen accompany Sater to meet with Mr. Artemenko? Who told Artemenko that these were the men with whom he should speak? The *Post* and *Times* reports broke within hours of each other on August 28, 2017. On the third of the same month, Andrew Rice quoted Sater as saying: "'In about the next 30 to 35 days,' he told me, 'I will be the most colorful character you have ever talked about. Unfortunately, I can't talk about it now, before it happens. And believe me, it ain't anything as small as whether or not they're gonna call me to the Senate committee.'" As Rice himself says, Sater made the statement to him before news of the June 16, 2016, meeting dropped, so the time lines don't match up. So we will wait and see whether Sater's words are merely his famous "salesmanship" or something more.[57]

Another Bayrock connection was Tamir Sapir, who hailed from Georgia (the country, not the state). Sapir, now deceased, was also an émigré from the Soviet Union, who came to the United States and "sold electronics to KGB agents from a storefront in Manhattan."[58] Tamir's son, Alex, was also involved in the Trump SoHo project. Alex Sapir's business partner and brother-in-law is Rotem Rosen, a former "right-hand man" of the Soviet-born Israeli billionaire Lev Leviev, an oligarch who boasts long-standing ties to Vladimir Putin and counts the Russian president as a "true friend."[59]

Leviev, who keeps a photo of the Russian president on a shelf in his office, figures into the Trump financial nexus in other ways. When Donald Trump was looking for real estate deals in Moscow, he met with Leviev in 2007 to discuss options.[60] Leviev also sold Jared Kushner a portion of the old *New York Times* building located in Times Square in October 2015, after Trump had announced running for president.[61]

Mortgages for the Trump SoHo building were issued by a Ukrainian businessman named Sam Kislin. Born in Odessa, Ukraine, in 1935, he emigrated to the United States in the 1970s. He eventually found his way to Brooklyn, New York, where, with the assistance of Russian émigrés, he started an electronics business. His partner in this venture was Tamir Sapir. In 1976, they sold two hundred televisions to Trump's Commodore Hotel.[62] He soon began trading with the Soviet Union and established a commodities trading business.[63] But the FBI has claimed that Kislin is a member of the Russian mob.[64] The International Police Organization (Interpol) in a 1996 report claimed that Kislin's firm, Trans Commodities, was used by two Uzbekistan mobsters for fraud and embezzlement.[65]

The Trump SoHo deal, among others like it, was done well

before Trump ever ran for office. But business relationships can be complicated. They can create alliances that lead to vulnerabilities. Will Trump feel an obligation to any former business partners? Will Trump's sons be in a position to profit off of partners looking for favors from the American president?

Trump's business deals have received enormous attention from the press.

Some argued that Trump was deeply corrupting the public square even before he took office, and presented a sanitized view of America's recent past. Professor Zephyr Teachout, who has done some very good work on the topic of corruption, bluntly argued, "Trump is upending 240 years of tradition and a core conviction of the founders: that a stable, safe, representative republic depends on protecting against the foreign corruption of our officeholders."[66]

Professor Teachout was no doubt unaware of the foreign corruption of officeholders as presented in this book concerning Chinese and Ukrainian commercial deals being done by the son of a vice president. Or the commercial ties between Senator McConnell's family and the Chinese government.

Teachout is correct in her observation that our constitutional republic is indeed at stake when politicians put family loyalty and estate building before loyalty to the country. Foreign corruption has been a longtime fear of America's leaders, going back to the Founding Fathers. Teachout quotes several founders at length, including Elbridge Gerry, who warned: "Persons having foreign attachments will be sent among us & insinuated into our councils, in order to be made instruments for their purposes. Everyone knows the vast sums laid out in Europe for secret services." Foreign commercial entanglements are a problem. But they didn't begin with Trump.[67]

Trump brings to office unique skills and complications as a businessman who has amassed properties and companies around the world. Initially, Trump dismissed the notion that the Trump Organization would be an ethical problem. I was one of those who called on Trump to disconnect himself from his businesses and place his assets in a blind trust. In the piece I cowrote in the *Washington Post* with Norm Eisen, I called on Trump to transfer "all his business interests to a blind trust or the equivalent." By doing so he "will set an important tone of integrity at the top for everyone in his administration as they address their own ethics and conflicts issues."[68]

It was not just a question of Trump doing all he could to separate himself from the family business. Trump's campaign promise to drain the swamp was an important one, and that message, I believe, was a key reason that he won the election. He needs to set a strong ethical example.

Some have argued that Trump needed to go further than just separating himself from the business. They argued that he needed to *sell* his holdings. If he still owned businesses like hotels, and foreign diplomats or businesses were paying to use those hotels, he would be violating the U.S. Constitution. This was a reference to the Emoluments Clause, which prevents U.S. officials from receiving payments or things of value from a foreign government or its agents.

It is worth noting that this issue never came up when Penny Pritzker was appointed commerce secretary. She, of course, would meet the definition of a U.S. official and she owned commercial real estate properties as well as hotels that did plenty of business with foreign governments and their agents. But more important, many legal scholars disagreed with this interpretation of the Emoluments Clause, arguing that the clause was meant to prevent gifts or bribes. To say that a president could not own a business that has some foreign clients was overly broad.

In the end, Trump did not quite set up a blind trust. His team of lawyers set up a trust whereby his children would run the business empire and he would remove himself from any decision making, but it was not "blind." Ethics experts say the scale of his holdings in the United States and around the world, and the close involvement of his children, together mean that the trust wouldn't extricate him from all potential conflicts of interest. *Salon* reported that Walter M. Shaub Jr., director of the Office of Government Ethics, in a speech at the Brookings Institution, "asserted that Trump 'stepping back from running his business is meaningless from a conflict of interest perspective,' and that 'limiting direct communication about the business is wholly inadequate . . . There's not supposed to be any information at all.'"[69]

In my experience, corrupt or questionable transactions can take time to uncover. In *Throw Them All Out*, I exposed insider trading on the stock market by members of Congress. The book was released in 2011; all of the examples of questionable transactions we uncovered occurred between 2008 and 2010. In *Extortion*, which was released in 2013, we found, among other things, the extortive fund-raising practices of members of Congress. Again, the examples we uncovered had occurred several years earlier. And *Clinton Cash*, which was released in 2015, was about donations and speaking fees the Clintons received between 2008 and 2012, during Hillary's tenure as secretary of state.

In his first eleven months in office, there have been thousands of articles and stories run by major media outlets on the commercial ties involving Trump and his adult children. Some reporting is solid, and some amounts to over-the-top punditry and poor reporting with anonymous sources. Either way, as aggravating as the latter is, the attention is good for the country. It means that people are watching.

However, the current level of media attention stands in sharp contrast to the lack of reporting by many of the same news outlets over the previous eight years on transactions involving Trump's predecessors or senior members on Capitol Hill.

This yawning gap is unhealthy. My frustration is *not* that the solid reporting on Trump has been *too tough* but that the reporting on the Obama administration was *way too soft* or in some cases *nonexistent.*

We can safely assume that if Donald Trump's children struck billion-dollar equity deals with the Chinese government, like Joe Biden's son did, it would make the front page of every newspaper across the country and receive a heavy rotation on the evening news.

So, where do we go from here?

CONCLUSION

- Corruption is not a victimless crime.
- America needs a Washington Corrupt Practices Act modeled after the Foreign Corrupt Practices Act.

The issue of political corruption is not new to our republic, even as it has evolved and changed over the course of the last two centuries. Today, corruption is exponentially more lucrative and complex. It is increasingly hard to enforce ethics standards that hold politicians to the spirit of the law and their sworn duty to serve their country, and prevent them from using public office for personal empire building. The mind-set that this is just the way the world works is itself corrupt thinking, which denies allegiance to the idea that America is an exceptional nation.

If we are to remain an effective constitutional republic, we must face and win the war on corruption at home. We must not tolerate public service as a front for family enrichment and elite will to power. It is un-American, and has direct and dire effects on policy-making and good governance.

Current ethics laws create a zero accountability zone for the Washington, D.C., political class in general. Let me illustrate this with two recent examples pulled from the headlines.

In China, the financial giant J. P. Morgan began a practice of hiring the children of government officials. They did this because they believed that it would enhance their ability to get more business in China. When news of the practice leaked, the federal government charged the company with violating the Foreign Corrupt Practices Act (FCPA). In short, the FCPA says that if you give someone something of value with the hope of getting special treatment from a foreign government, you are committing a corrupt act and can be charged and prosecuted. J. P. Morgan settled with the federal government to the tune of $264 million.[1]

Now imagine if J. P. Morgan had done this not in Shanghai, but in Washington, D.C. Nothing would have happened. It would be completely legal. In fact, as we have seen, special deals for the children of American politicians and the practice of hiring the children of politicians to serve as lobbyists is commonplace.

Consider another recent corruption case.

Governor Bob McDonnell of Virginia and his wife, Maureen, were given $135,000 in gifts, loans, and trips by a businessman named Jonnie Williams Sr. while they occupied the governor's mansion in Richmond. The couple received, among other things, a Rolex watch, shopping trips in New York, and clothes. Williams, who was the CEO of Star Scientific, a producer of dietary supplements, was hoping that the state of Virginia might review his product and vouch for it. McDonnell was charged with public corruption and convicted. But federal courts later overturned the conviction, based on existing law.[2]

Now, imagine if Jonnie Williams had given those gifts to a governor in China. Had it been similiarly exposed, it would have been a clear violation of the FCPA.

The question then is: Why do we have a Foreign Corrupt Practices Act but not a Washington Corrupt Practices Act? Or, put a different way, why are U.S. lawmakers (and governors, for that matter, through robust state laws) themselves not held accountable?

A Washington Corrupt Practices Act would be a first step and could be based largely on the language in the FCPA. By strengthening U.S. code on giving gifts, favors, or deals to politicians or their families in the hopes of getting favors in return—which is the very essence of bribery—we would deal dramatically with the problem of America's Princelings. Right now, U.S. law is harsher on corrupt deals involving Chinese Princelings than it is on those involving American Princelings. That is not a good place for our country to be.

Another necessary reform: we need to broaden financial disclosures by American politicians. Transparency really is the best disinfectant. One of the reasons we are seeing the rise of American Princelings is because the adult children of American politicians do not have to disclose their financial transactions or assets.

I recognize that it might seem intrusive to expect the adult children of politicians to disclose publicly exactly what they are earning. After all, they usually do not choose to have parents who hold public office.

To strike a balance between disclosure and privacy, the adult children of America's political leaders could at least be required to disclose transactions with foreign entities above, say, $150,000 a year. If they are working for an American corporation on a career track, we don't need to know their salaries, but if they are involved in large financial transactions with foreign entities, we do.

Some might argue that this is unfair. But the fact of the matter is that holding public office is a privilege; it is not a right. It comes with unique burdens for oneself and one's family, much as it does for those in military service. The goal of such ethics laws is to protect the Constitution—the pact between the elected and the electorate—that makes us a republic and not an oligarchy.

The third necessary reform relates to the phenomenon of

political intelligence. After the publication of my book *Throw Them All Out*, Congress passed, and President Obama signed, the Stop Trading on Confidential Information (STOCK) Act, which made insider trading illegal for government officials. It was an important reform, which was unfortunately gutted a few months later by the same Congress and president, once the public spotlight moved on to other issues.

It is great to have a law that makes insider trading illegal, but it is hard to enforce. The law is legally easy to circumvent: "political intelligence" is big business in Washington and basically amounts to the selling of information on what actions the government is going to take that might affect a company. We can tighten the STOCK Act to deal explicitly with the smash and grab problem. As with corruption in general, smash and grab is not a victimless crime. "Smashing" a company throws thousands of people out of work and disrupts the lives of many, so that someone else can "grab" and profit. Only by changing existing laws can we deter this kind of grossly unethical behavior.

Finally, a word must be said to politicians themselves. Unfortunately, many politicians in Washington believe they are entitled to on-the-job enrichment and building family empires at our expense. Some in the nation's capital want to do the right thing. They must make clear to their children that they will be offered sweetheart deals and that they must not take them. Back in May 1988, then vice president George H. W. Bush was running for president. He sent a letter to his eldest son, George W. Bush. "We are about to sail into unchartered waters, in terms of family scrutiny," he wrote. "We've all been through a lot of inquiry and microscopic probing; however, it'll get worse, not just for our family."[3]

Father advised son: "As we move closer to November, you'll find you've got a lot of new friends. They may become real friends . . . My plea is this: please do not contact any federal agency or department on anything. A call from a 'Bush' will get returned, but there is a great likelihood that it will be leaked; maybe deliberately misrepresented.'"[4]

Ethical reputation is a legacy worth passing along, to one's own children, and one's country.

ACKNOWLEDGMENTS

Writing a book on a complex and important subject such as this requires a talented team to bring to successful completion. This book represents the culmination of many hours of diligent work by a diverse group of patriotic people.

I was blessed to have the assistance of the following researchers on this project: Brian Baugus, Daniel Boyle, David Brown, Seamus Bruner, Keegan Connolly, Jacob McLeod, Steve Stewart, and Peggy Sukhia. Thanks also go to interns Christina Armes and Hannah Cooperman. I also appreciate the help of Matthew Tyrmand, who helped us connect with researchers in Ukraine for that portion of this project, and David Lawrence, who offered keen legal insight into the remedies proposed in this book.

I cofounded the Government Accountability Institute (GAI) in 2012 with Stephen K. Bannon, who stepped down from the board in 2016 to pursue other noteable goals. Today the organization enjoys the benefits of a tremendous board of directors including Chairman Rebekah Mercer, who has enthusiastically supported our work and provided thoughtful leadership. Thanks also go to board members Hunter Lewis, Ron Robinson, and Owen Smith for their insightful guidance.

At GAI we have a formidable band of professional staff who make us one of the most effective investigative organizations in the world. These include Stuart Christmas our general counsel, research director Eric Eggers, program director Steve Post, and data and technical director Chance Hendrix. Sandy Schulz helps us expertly navigate the world of media and communications. They all contributed mightily to this book. Some GAI employees who also contributed have asked to remain anonymous.

Over the course of the past ten years, I have benefited tremendously from my agents, Glen Hartley and Lynn Chu. I thank them for their continued wise counsel. This is my first book with editor Eric Nelson. Thanks, Eric, for your keen insights and helpful suggestions. You made the long editing process relatively painless.

Finally, thanks to my wonderful wife, Rhonda, and to my entire family. Your encouragement and support over the years are much appreciated. I have dedicated this book to my son, Jack, and my daughter, Hannah.

The author alone is responsible for the contents of this book.

NOTES

CHAPTER 1: CORRUPTION BY PROXY

1 G. Calvin Mackenzie, *Scandal Proof: Do Ethics Laws Make Government Ethical?* (Washington, D.C.: Brookings Institution Press, 2002), 19, https://books.google.com/books?id=C5Y4Ws9Ut4MC&pg=PA19&lpg=PA19&dq=%22Public+office+is+a+privilege,+not+a+right,+and+people+who+accept+the+privilege+of+holding+office+in+the+government+must+of+necessity+accept+that+their+entire+conduct+should+be+open+to+inspection+by+the+people+they+are+serving.%22+truman&source=bl&ots=B7YXNhrNsh&sig=9NxASjhGedU6rVTdzwE8pkw5rpI&hl=en&sa=X&ved=0ahUKEwj1pc3ClKrVAhWFNiYKHVD_AjkQ6AEIKDAA#v=onepage&q=%22Public%20office%20is%20a%20privilege%2C%20not%20a%20right%2C%20and%20people%20who%20accept%20the%20privilege%20of%20holding%20office%20in%20the%20government%20must%20of%20necessity%20accept%20that%20their%20entire%20conduct%20should%20be%20open%20to%20inspection%20by%20the%20people%20they%20are%20serving.%22%20truman&f=false.

2 "75% in U.S. See Widespread Government Corruption," *Gallup*, September 19, 2015, http://www.gallup.com/poll/185759/widespread-government-corruption.aspx.

3 "Beyond Distrust: How Americans View Their Government," Pew Research Center, November 23, 2015, http://www.people-press.org/2015/11/23/beyond-distrust-how-americans-view-their-government/.

4 Jerry Markon, "Ex-Rep. Jefferson (D-La.) Gets 13 Years in Freezer Cash Case," *Washington Post*, November 14, 2009, http://www.washingtonpost.com/wp-dyn/content/article/2009/11/13/AR2009111301266.html.

5 Charles R. Babcock and Jonathan Weisman, "Congressman Admits Taking Bribes, Resigns," *Washington Post*, November 29, 2005, http://www.washingtonpost.com/wp-dyn/content/article/2005/11/28/AR2005112801827.html.

6 Neil A. Lewis, "Abramoff Gets 4 Years in Prison for Corruption," *New York Times*, September 4, 2008, http://www.nytimes.com/2008/09/05/washington/05abramoff.html.

7 Susan Schmidt and James V. Grimaldi, "Ney Pleads Guilty to Corruption Charges," *Washington Post*, October 14, 2006, http://www.washingtonpost.com/wp-dyn/content/article/2006/10/13/AR2006101300169.html.

8 "Taking Stock of Congress," Stanford Alumni website, November/December 2012, https://alumni.stanford.edu/get/page/magazine/article/?article_id=57554.

9 Peter Schweizer, "Throw Fear," in Schweizer, *Extortion* (New York: Houghton Mifflin, 2013), 19.

10 Peter Schweizer, "Quid Pro Quo?" in Schweizer, *Clinton Cash* (New York: HarperCollins, 2015), 180.

11 Dana Blanton, "Fox News Poll: Clinton Up by 6 Points, 89 Percent Say 'Hot-headed' Describes Trump," Fox News, June 29, 2016, http://www.foxnews.com/politics/2016/06/29/fox-news-poll-clinton-up-by-6-points-89-percent-say-hot-headed-describes-trump.html; "Clinton's Trustworthiness Remains a Drag on Her Candidacy," *Rollcall*, August 11, 2016, https://www.rollcall.com/politics/clintons-trustworthiness-remains-a-drag-on-her-candidacy.

12 "Secret Recordings Fueled FBI Feud in Clinton Probe," *Wall Street Journal*, November 2, 2016, https://www.wsj.com/articles/secret-recordings-fueled-fbi-feud-in-clinton-probe-1478135518.

13 Allen Gannett and Chad Rector, "The Rationalization of Political Corruption," *Public Integrity* 17, no. 2 (March 2015): 165–166, EBSCOhost.

14 Laurence Cockcroft, "What Is Corruption?" in Cockcroft, *Global Corruption: Money, Power and Ethics in the Modern World* (London: I. B. Tauris, 2012), 3.

15 William L. Riordon, *Plunkitt of Tammany Hall: A Series of Very Plain Talks on Very Practical Politics* (New York: Signet, 2015).

16 Scott J. Basinger, "Scandals and Ethics Reform in the U.S. House of Representatives," *Public Integrity* 18, no. 4 (June 2016): 360, EBSCOhost.

17 *Annual Report on Activities Undertaken in 2012*, OECD Working Group on Bribery, https://www.oecd.org/daf/anti-bribery/AntiBriberyAnnRep2012.pdf, 6.

18 Robert C. Brooks, "The Nature of Political Corruption," *Political Science Quarterly* 24, no. 1 (March 1909): 15, http://www.jstor.org/stable/pdf/2141078.pdf?refreqid=excelsior%3Ae908eeafccf34dc6d2d68634faaf079e.

19 Ibid.

20 Marilee Hanson, "Mary, Queen of Scots: Biography, Facts, Portraits & Information," *English History*, January 31, 2015, https://englishhistory.net/tudor/relative/mary-queen-of-scots/.

21 Stephen Hess, "The Best Butter," in Hess, *America's Political Dynasties* (New Brunswick, NJ: Transaction, 1997), 4.

22 Ibid., 5.

23 "The Billy Carter-Libyan Connection: A Chronology," *Christian Science Monitor*, August 1, 1980, http://www.csmonitor.com/1980/0801/080150.html.

24 Committee on Government Reform, *Justice Undone: Clemency Decisions in the Clinton White House*, H.R. Rep. No. 107–454, at 717, vol. 1 (Washington, DC: U.S. Government Printing Office, 2002), https://www.congress.gov/107/crpt/hrpt454/CRPT-107hrpt454-vol1.pdf.

25 "Bush Brother's Divorce Reveals Sex Romps," *CNN.com*, November 25, 2003, http://www.cnn.com/2003/ALLPOLITICS/11/25/bush.brother.reut/.

26 David Barboza, "Billions in Hidden Riches for Family of Chinese Leader," *New York Times*, October 25, 2012, http://www.nytimes.com/2012/10/26/business/global/family-of-wen-jiabao-holds-a-hidden-fortune-in-china.html.

27 Huhua Cao and Vivienne Poy, eds., *The China Challenge: Sino-Canadian Relations in the 21st Century* (Ottawa, ON: University of Ottawa Press, 2011), 72.

28 Peter Armstrong, "CSIS Alleges Canadian Politicians Under the Influence of a Foreign Government," CBC Radio, June 23, 2010, EBSCOhost; "China Swaying Politicians with Gifts, Sex: MP," CBC News, July 30, 2010, http://www.cbc.ca/news/politics/china-swaying-politicians-with-gifts-sex-mp-1.891918.

29 Sarah Chayes, "Remedies," in Chayes, *Thieves of State: Why Corruption Threatens Global Security* (New York: W. W. Norton, 2015), 185.

30 Samuel P. Huntington, "The Crisis of National Identity," in Huntington, *Who Are We? The Challenges to America's National Identity* (New York: Simon & Schuster, 2004), 7.

31 Ibid., 287.

32 Dolia Estevez, "Mexico Is Among the Top 10 Countries Paying Washington Lobbyists," *Forbes*, May 29, 2014, https://www.forbes.com/sites/doliaestevez/2014/05/29/mexico-is-among-the-top-10-countries-paying-washington-lobbyists/#1f865dfb1416.

33 Cockcroft, "Why Now?" in Cockcroft, *Global Corruption*, 103.

34 Paul Harvey, *Remember These Things* (Chicago: Heritage Foundation, 1952), 57.

35 Basinger, "Scandals and Ethics Reform."

36 As quoted in Cockcroft, "Constant Values, Changing Standards," in Cockcroft, *Global Corruption*, 93.

37 Niccolò Machiavelli, *The Prince*, trans. (into French) Marie Gaille-Nikokimov (Paris: Librairie générale française, 2000), chap. 19, p. 131, quoted in Chayes, "Lord King, How I Wish That You Were Wise," in Chayes, *Thieves of State*, 9.

CHAPTER 2: AMERICAN PRINCELINGS: TWO SONS AND A ROOMMATE

1 Jules Witcover, *Joe Biden: A Life of Trial and Redemption* (New York: William Morrow/HarperCollins, 2010), 356, 360.

2 "Rosemont Seneca Partners, LLC," opencorporates.com, accessed February 17, 2017, https://opencorporates.com/companies/us_de/4703140.

3 "Marianna Fazylova, CAIA Vice President," biography, ssiinvest.com, accessed February 22, 2017, www.ssiinvest.com/page55.html.

4 Christopher Drew and Mike McIntire, "Obama Aides Defend Bank's Pay to Biden Son," *New York Times*, August 24, 2008, http://www.nytimes.com/2008/08/25/us/politics/25biden.html?mcubz=1.

5 Christopher Drew, "Campaign Says Biden Son Dropped Lobbying Clients," *New York Times*, September 13, 2008, http://www.nytimes.com/2008/09/13/us/politics/13resign.html.

6 Open Secrets, "Oldaker, Biden & Belair," Total Lobbying Income 2008, accessed February 14, 2017, https://www.opensecrets.org/lobby/firmsum.php?id=D000036374&year=2008.

7 Drew, "Campaign Says Biden Son Dropped Lobbying Clients."

8 Lottito v. Biden, No. 07/60045, James B. Biden Affidavit (New York Supreme Court, April 13, 2007), https://www.scribd.com/doc/14758437/James-Biden-sworn-statement.

9 Sam Jones, Tracy Alloway, and Stacy-Marie Ishmael, "The Politics of Paradigm," *Financial Times*, May 1, 2009, https://ftalphaville.ft.com/2009/05/01/55288/the-politics-of-paradigm/.

10 Lottito v. Biden, No. 07/60045, James Biden Affidavit.

11 Ben Smith, "Biden Son's, Brothers Firm Downplays Ties to Closed Hedge Fund," *Politico*, April 29, 2009, www.politico.com/blogs/ben-smith/2009/04/biden-sons-brothers-firm-downplays-ties-to-closed-hedge-fund-updated-017947; "SEC Charges Connecticut-Based Hedge Fund in Multi-Million Dollar Fraud," U.S. Securities and Exchange Commission, April 27, 2009, https://www.sec.gov/news/press/2009/2009–90.htm.

12 "SEC Charges Connecticut-Based Hedge Fund in Multi-Million Dollar Fraud."

13 Susan Schmidt, Steve Stecklow, and John R. Emshwiller, "Stanford Had Links to a Fund Run by Bidens," *Wall Street Journal*, February 24, 2009, https://www.wsj.com/articles/SB123543815326954907; Nathan Vardi, "Allen Stanford Convicted in $7 Billion Ponzi Scheme," *Forbes*, March 6, 2012, http://www.forbes.com/sites/nathanvardi/2012/03/06/allen-stanford-convicted-in-7-billion-ponzi-scheme/#48377e342b08.

14 Curtis Wilkie, *The Fall of the House of Zeus: The Rise and Ruin of America's Most Powerful Trial Lawyer* (New York: Crown, 2010), 195.

15 Witcover, *Joe Biden*, 388.

16 The Harvard Salient, October 27, 2008, citing *Washington Times*, October 15, 2008, http://www.washingtontimes.com/news/2008/oct/15/biden-routes-campaign-cash-to-family-their-firms/.

17 Ryan D'Agostino, "Things My Father Taught Me: An Interview with Joe and Hunter Biden," *Popular Mechanics*, May 18, 2016, http://www.popularmechanics.com/home/a20655/things-my-father-taught-me/.

18 "Ted Kennedy Makes Final Trip to Washington for Burial at Arlington Cemetery," *Zimbio*, photograph, August 28, 2009, http://www.zimbio.com/pictures/007gqypJhpa/Ted+Kennedy+Makes+Final+Trip+Washington+Burial/LfsPaxUauCQ/Hunter+Biden.

19 David Jackson, "Obama's Guests at India State Dinner," *USA Today*, November 24, 2009, http://content.usatoday.com/communities/theoval/post/2009/11/obamas-guests-at-india-state-dinner/1#.WKR5ohCXK3Z.

20 Mitchell Layton, "Duke v Georgetown," *Zimbio*, photograph, January 29, 2010, http://www.zimbio.com/photos/Barack+Obama/Hunter+Biden.

21 Moses Robinson, "Usher's New Look Foundation—World Leadership Conference & Awards 2011—Day 3," *Zimbio*, photograph, July 22, 2011, http://www.zimbio.com/photos/Hunter+Biden/Usher+New+Look+Foundation+World+Leadership/En-xEe0qMgo.

22 Joyce Gannon, "Pittsburgh's Charms Attract Another Young Professional: Christopher Heinz," *Pittsburgh Post-Gazette*, November 29, 2015, http://www.post-gazette.com/local/city/2015/11/29/Pittsburgh-s-charms-lure-another-young-professional-Christopher-Heinz/stories/201511290063.

23 "Weddings; Teresa Heinz and John F. Kerry," *New York Times*, May 28, 1995, http://www.nytimes.com/1995/05/28/style/weddings-teresa-heinz-and-john-f-kerry.html.

24 "Christopher D. Heinz," Our Campaigns, accessed February 15, 2016, http://www.ourcampaigns.com/CandidateDetail.html?CandidateID=47631.

25 "Rosemont Seneca Partners Leadership Team," accessed on Wayback Machine on February 15, 2016, http://web.archive.org/web/20100704182256/http:/www.rosemontseneca.com/team.html.

26 Anna Schneider-Mayerson, "Baby Bundlers Feed Democrats," *Observer*, August 2, 2004, http://observer.com/2004/08/baby-bundlers-feed-democrats/.

27 Ibid.

28 Mollie Hemingway, "9 Questions to Ask About Biden's Work with a Gas Company in Ukraine," *Federalist*, May 13, 2014, http://thefederalist.com/2014/05/13/9-questions-to-ask-about-bidens-work-with-a-gas-company-in-ukraine/; Ralph Vartabedian, "Estimate of Heinz Fortune Doubled," *Los Angeles Times*, June 27, 2004, http://articles.latimes.com/2004/jun/27/nation/na-heinz27.

29 Darlene Superville, "Michelle Obama Welcomes G-20 Spouses, Hosts Dinner at Teresa Heinz Kerry's Rosemont Farm," *Huffington Post*, June 27, 2010, http://www.huffingtonpost.com/2009/09/25/michelle-obama-welcomes-g_n_299644.html.

30 Jason Galanis, e-mail message to Michelle Morton, June 3, 2014, "Introducing COR Capital Summary" (unpublished manuscript), Microsoft Word file, accessed on February 18, 2017, http://www.bancexposed.com/wp-content/uploads/2016/10/Galanis-COR-Email-1.pdf. They also took an

advertisement out in a legal publication in China stating that they had $2 billion under management. "渤 海产业投资基金招聘法务实习生," job solicitation, accessed February 18, 2017, http://lawtalks.cn/jobs/wap/index-wap2.php?p=2135.

31 "Chris Heinz [Derived Headline]," article excerpt from *Pittsburgh Tribune-Review*, n.d., Questia, https://www.questia.com/read/1P2–39141495/chris-heinz-derived-headline.

32 "Rosemont Seneca Partners," MapQuest location, accessed February 15, 2017, https://www.mapquest.com/us/district-of-columbia/business-washington/rosemont-seneca-partners-44193124.

33 "China Profile—Full Overview," *BBC News*, January 21, 2016, http://www.bbc.com/news/world-asia-pacific-13017879.

34 Steve Clemons, "The Geopolitical Therapist," *Atlantic*, August 26, 2016, https://www.theatlantic.com/international/archive/2016/08/joe-biden-interview/497633/.

35 Sarah Schafer, "China's Family Ties," *Newsweek*, October 6, 2002, http://www.newsweek.com/chinas-family-ties-146669.

36 David J. Lynch, Jennifer Hughes, and Martin Arnold, "JPMorgan to Pay $264m Penalty for Hiring 'Princelings,'" *Financial Times*, November 17, 2016, https://www.ft.com/content/fc32b64e-ac87–11e6-ba7d-76378e4fef24.

37 Anthony Goh and Matthew Sullivan, "The Most Misunderstood Business Concept in China," *Business Insider*, February 24, 2011, http://www.businessinsider.com/the-most-misunderstood-business-concept-in-china-2011–2.

38 "Thornton Group to Bring Rosemont Seneca High-Level Access to China's Financial/Fund Industry Executives," *Thornton Company News*, April 12, 2010, https://translate.google.com/translate?hl=en&sl=zh-CN&u=http://www.thorntonasia.com/chinese/newscount.asp%3FArticleID%3D282&prev=search.

39 Amy Padnani and Katharine Q. Seelye, "Whitey Bulger: The Capture of a Legend," *New York Times*, n.d., http://www.nytimes.com/interactive/us/bulger-timeline.html?_r=0#/#time256_7542.

40 "Company Overview of Thornton Group LLC," *Bloomberg*, October 1, 2017, http://www.bloomberg.com/research/stocks/private/people.asp?privcapId=108413293.

41 Katharine Q. Seelye, "Sticking by a Murderous Brother, and Paying for It Dearly," *New York Times*, November 24, 2013, http://www.nytimes.com/2013/11/25/us/sticking-by-a-murderous-brother-and-paying-for-it-dearly.html; Howie Carr, "'Whitey' Bulger, the Alcatraz-Hardened King of the Boston Underworld, Begins Murder Trial Next Week," *New York Daily News*, June 9, 2013, http://www.nydailynews.com/news/crime/whitey-bulger-king-boston-underworld-murder-trial-article-1.1367313.

42 "Thornton Group to Bring Rosemont Seneca High-Level Access."

43 Ibid. Among the pictured are Gao Xiqing, general manager of the China Investment Co., the Chinese government's sovereign wealth fund, which manages hundreds of billions in Chinese foreign exchange reserves; Ji Guogzhong, of the National Council for Social Security Fund; Cui Yong, China Life Assets Management Company vice president; Peng Zuogang, the general manager of the China Postal Savings Bank, the second-largest retail bank in China, with forty thousand branches around the country; and Wei Xin, chairman of the Founder Group, which handles investments for the government's Peking University, and others.

44 Ibid.; "Nuclear Security Summit Kicks Off in Washington Hu Jintao Delivers an Important Speech," Embassy of the People's Republic of China in the United States of America, April 14, 2010, http://www.china-embassy.org/eng/zt/aqfh/t679665.htm.

45 "Thornton Met with Taiwan's Financial Holding Senior," *Thornton Company News*, May 24, 2011, http://www.thorntonasia.com/chinese/newscount.asp?ArticleID=288.

46 "Remarks by Vice President Joe Biden to the Opening Session of the U.S.-China Strategic and Economic Dialogue," Briefing Room, The White House Office of the Vice President, May 9, 2011, https://obamawhitehouse.archives.gov/the-press-office/2011/05/09/remarks-vice-president-joe-biden-opening-session-us-china-strategic-econ.

47 Ng Han Guan, "US Vice President Joe Biden Waves as He Walks Out of Air Force Two," Getty Images, December 4, 2013, accessed February 15, 2017, http://www.gettyimages.com/detail/news-photo/vice-president-joe-biden-waves-as-he-walks-out-of-air-force-news-photo/453377437#vice-president-joe-biden-waves-as-he-walks-out-of-air-force-two-with-picture-id453377437.

48 Steve Holland and Paul Eckert, "Biden to Emphasize Asia Pivot on Trip to China, Japan, South Korea," Reuters, November 4, 2013, http://www.reuters.com/article/us-usa-biden-china-idUSBRE9A30C520131104.

49 Steve Clemons, "Biden's 330-Minute Balancing Act in China," *Atlantic*, December 4, 2013, https://www.theatlantic.com/china/archive/2013/12/biden-s-330-minute-balancing-act-in-china/282061/.

50 Mariano Castillo and Jethro Mullen, "In Japan, Biden Promises to Press China on Airspace Dispute," *CNN.com*, December 3, 2013, http://www.cnn.com/2013/12/03/world/asia/biden-asia-trip/index.html.

51 Tim Kelly and Phil Stewart, "Defying China, U.S. Bombers and Japanese Planes Fly Through New Air Zone," Reuters, November 27, 2013, http://www.reuters.com/article/us-china-defense-usa-idUSBRE9AP0X320131127.

52 Mark Landler and Martin Fackler, "Biden Walks a Fine Line in Japan; He Voices Concern About China's Air Zone, but No Call for a Reversal," *Inter-*

national New York Times, December 4, 2013, Questia; Castillo and Mullen, "In Japan, Biden Promises to Press China."

53 Associated Press Archive, "China Biden 2," December 4, 2013, footage, 00:54, Joe Biden's trip to Beijing, China, http://www.aparchive.com/metadata/youtube/111ef68664df89c93e9b20fa8a6b47d7.

54 Mark Landler, "Biden Tries to Soothe Asia Tension; Political Instincts Come in Handy While Tackling Sensitive Issues in Region," *International New York Times*, December 7, 2013, Questia, https://www.questia.com.newspaper/1P2–36312038/biden-tries-to-soothe-asia-tension-political-instincts; Matt Schiavenza, "White House Sends Joe Biden to China to Cover *The Atlantic*'s Steve Clemons," *Atlantic*, December 5, 2013, http://www.theatlantic.com/china/archive/2013/12/white-house-sends-joe-biden-to-china-to-cover-em-the-atlantic-em-s-steve-clemons/282071.

55 Clemons, "Biden's 330-Minute Balancing Act." Note: Joe Biden was familiar with dealing with senior Chinese officials. He had long established ties in China, having first visited the country as a U.S. senator in 1979 when he met Deng Xiaoping. As a member of the Senate Intelligence Committee, the young senator had created waves by suggesting that China allow the United States to build two American spy installations on the Chinese-Soviet border. Many visits to China over the course of more than two decades followed. In addition, he often met with Chinese delegations visiting Washington. He maintained a highly personal approach to diplomacy. Vice President Biden and President Xi considered the other a friend. They had met previously in Chengdu, China, and in Los Angeles for highly productive and personal talks.

56 Jacqueline Newmyer Deal, "Red Alert," *The National Interest*, May/June 2014, 85–96.

57 Jane Morse, "Biden Urges 'Practical Cooperation' in U.S.-China Relations," *IIP Digital*, December 5, 2013, http://iipdigital.usembassy.gov/st/english/article/2013/12/20131205288461.html#axzz2mvweJcB3.

58 "Background Press Briefing on Vice President Biden's Trip to China, Japan and the Republic of Korea," The White House Office of the Press Secretary, November 27, 2013, https://obamawhitehouse.archives.gov/the-press-office/2013/11/27/background-press-briefing-vice-president-bidens-trip-china-japan-and-rep.

59 David Kan Ting, "China Abandons Deng's Teachings," *China Post*, December 13, 2013.

60 "Remarks by the Vice President at a Breakfast with the American Chamber of Commerce in Beijing and the U.S.-China Business Council," The White House Office of the Vice President, December 5, 2013, https://obamawhitehouse.archives.gov/the-press-office/2013/12/05/remarks-vice-president-breakfast-american-chamber-commerce-beijing-and-u.

61 Schiavenza, "White House Sends Joe Biden to China to Cover *The Atlantic*'s Steve Clemons."
62 Leslie Larson, "Joe Biden's 15-year-old Granddaughter, Finnegan, Gets a Turn Traveling Asia with 'Pop' Biden," *New York Daily News*, December 6, 2013, http://www.nydailynews.com/news/politics/biden-granddaughter-joins-vp-asia-trip-article-1.1539770.
63 Schiavenza, "White House Sends Joe Biden to China to Cover *The Atlantic*'s Steve Clemons."
64 "Bohai Huamei (Shanghai) Equity Investment Fund Management Co., Ltd.," www.pedata.cn, accessed on Google Translate, December 16, 2013, https://translate.google.com/translate?hl=en&sl=zh-CN&u=http://org.pedata.cn/21180219.html&prev=search.
65 James Reilly, "China's Economic Statecraft: Turning Wealth into Power," Lowy Institute for International Policy Analysis, November 27, 2013.
66 Li-Wen Lin, "A Network Anatomy of Chinese State-Owned Enterprises" (European University Institute Working Paper RSCAS, July 2017).
67 "Tian Guoli," China Vitae, accessed February 21, 2017, http://www.chinavitae.com/biography/Tian_Guoli.
68 Joe McDonald, "China's Xi Expresses Confidence in US Economic Strength as He and Biden Meet Business Leaders," *Canadian Press*, August 19, 2011.
69 *Encyclopedia Britannica Online*, s.v. "Bo Hai," accessed February 21, 2017, http://original.britannica.com/eb/article-9067596.
70 "About Us—Overview," BHR Partners website, accessed February 21, 2017, http://en.bhrpe.com/_d276561105.htm.
71 "About Us—Strengths," BHR Partners website, accessed February 21, 2017, http://en.bhrpe.com/_d276561113.htm.
72 "Bohai Fund to Take Advantage of Home-Grown Status," U.S. Department of State, Beijing, China, *WikiLeaks*, January 17, 2008, https://wikileaks.org/plusd/cables/08BEIJING164_a.html.
73 Tim Burroughs, "Crest of a Wave: Industry Q&A: Jonathan Li and Xin Wang," *Asian Venture Capital Journal*, March 1, 2016, 19.
74 "BHR Newsletter," June 2015, http://en.bhrpe.com/newsitem/276845997.
75 "About Us—Overview," BHR Partners website, accessed February 21, 2017.
76 "The Next Shenzhen?" *Economist*, October 3, 2013, http://www.economist.com/news/china/21587237-new-enterprise-zone-could-spark-wider-market-reformsbut-only-if-bureaucrats-ease-their-grip.
77 Ely Ratner, "Course Correction: How to Stop China's Maritime Advance," *Foreign Affairs*, July/August 2017, 65.
78 Simon Denyer, "China's Rise and Asian Tensions Send U.S. Relations into Downward Spiral," *Washington Post*, July 7, 2014, https://www.washingtonpost.com/world/asia_pacific/chinas-rise-and-asian-tensions-send-us-relations-into-downward-spiral/2014/07/07/f371cfaa-d5cd-4dd2–925c-246c099f04ed_story.html?utm_term=.0c5e285600ca.

79 Ibid.; Li Jiabao, "China-US Investment Treaty on Fast Track," *ChinaDaily.com*, July 10, 2014, http://www.chinadaily.com.cn/world/2014–07/10/content_17696571.htm; Jeremy Page and Ian Talley, "U.S. Seeks to Salvage Dialogue with China at Beijing Summit," *Wall Street Journal*, July 7, 2014, https://www.wsj.com/articles/u-s-seeks-to-salvage-dialogue-with-china-at-beijing-summit-1404752026.

80 Stephen Shaver, "Kerry, Lew and Baucus Visit the Great Wall Outside of Beijing," UPI, July 8, 2014, http://www.upi.com/News_Photos/view/upi/ac6ae8f8c5bdcffebb5d3b06042dc71c/Kerrry-Lew-and-Baucus-visit-the-Great-Wall-outside-of-Beijing/.

81 Martin Sieff, "John Kerry's moment in Asia," Chinadaily.com, April 12, 2013, http://usa.chinadaily.com.cn/opinion/2013-04/12/content_16395184.htm.

82 Li Jiabao, "China-US Investment Treaty on Fast Track."

83 Sarah Larimer, "John Kerry's Guitar Diplomacy: A Critical Review (updated with Esteban's Thoughts)," *Washington Post*, July 10, 2014, https://www.washingtonpost.com/blogs/in-the-loop/wp/2014/07/10/john-kerrys-guitar-diplomacy-a-critical-review/?utm_term=.e4eba099f8b8; Reuters, "Kerry Plays Classical Guitar in China," *YouTube*, July 10, 2014, https://www.youtube.com/watch?v=QE6YHyVM8R0; "File: John Kerry Playing Guitar in China 2014.jpg," *Wikimedia Commons*, July 10, 2014, https://commons.wikimedia.org/wiki/File:John_Kerry_playing_guitar_in_China_2014.jpg.

84 Associated Press, "US and China Had Frank Cyber-Hacking Talks: John Kerry," Gadgets360, July 10, 2014, https://gadgets.ndtv.com/internet/news/us-and-china-had-frank-cyber-hacking-talks-john-kerry-556328.

85 "U.S.-China Economic Alliance," Editorials, *Wikimedia Commons*, July 19, 2014, https://editorials.voa.gov/a/us-china-economic-alliance/1962287.html.

86 Note: the article dated August 2015 about the sale of Rosemont says that the negotiations began more than a year earlier, meaning July 2014 at the latest: Bruce Krasnow, "Global Development Firm Purchases Rosemont Realty," *Santa Fe New Mexican*, August 24, 2015, http://www.santafenewmexican.com/news/business/global-development-firm-purchases-rosemont-realty/article_2fa9787e-350d-5d44–83e3-b24b1300978f.html.

87 Rosemont Realty, according to the company's internal documents, "operates as a subsidiary of Rosemont Solebury Capital Management LLC." And who runs Rosemont Solebury? That would be Christopher Heinz, the managing director, stepson of John Kerry, and Devon Archer, the cofounder and chief operating officer. It is a subsidiary of Rosemont Capital, which is directly tied to the Heinz Family Office. In corporate documents, it is clear that Rosemont Capital and its subsidiary Rosemont Realty were funded by the Heinz family fortune. In several corporate biographies, Kerry confidant

Devon Archer describes himself as the "Co-Founder and Managing Partner, Rosemont Capital, LLC (Heinz Family Office)." The Heinz Family Office, of course, is the wealth center of the Kerry-Heinz family, managing their financial and other holdings.

88 "Board of Directors," Burisma, n.d., http://archive.is/Ia4q4.

89 "Rosemont Realty," Scribd, n.d., https://www.scribd.com/document/165097602/Rosemont.

90 Gail Schontzler, "Who Is Dan Burrell, Businessman Behind Controversial Doctor College?" *Bozeman Daily Chronicle*, November 30, 2015, http://www.bozemandailychronicle.com/news/education/who-is-dan-burrell-businessman-behind-controversial-doctor-college/article_bb325702-b8f9–5f09-bdc7–193bd22b210e.html.

91 Anne-Marie O'Connor, "Kerry's California Coddling," *Los Angeles Times*, October 6, 2004, http://articles.latimes.com/2004/oct/26/local/me-dems26/3.

92 Krasnow, "Global Development Firm Purchases Rosemont Realty."

93 Schontzler, "Who Is Dan Burrell, Businessman Behind Controversial Doctor College?"

94 Krasnow, "Global Development Firm Purchases Rosemont Realty."

95 Ibid.

96 See the entry for Mr. Li Ming on "Company Board," n.d., Sino-Ocean website, accessed October 10, 2017, http://www.sinooceanland.com/en-US/Investor/CompanyBoard.

97 Krasnow, "Global Development Firm Purchases Rosemont Realty."

98 This is from a bio that was discovered in the SEC Edgar archives. https://www.sec.gov/Archives/edgar/data/1169770/000089706916001088/cg816992.pdf, p. 17.

99 Sino-Ocean Land Holdings Limited, *Annual Report 2013*, http://www.finanznachrichten.de/pdf/20140404_085651_NULL_P4MF1E1L0SLZY4YS.1.pdf.

100 The chairman of Sino Ocean Land is also the chairman of COSCO. "Members of Board of Directors," CIMC, n.d., http://web.cimc.com/en/ir/governance/board/201001/t20100129_5074.shtml.

101 "Company Profile," Sino-Ocean Land, n.d., http://www.sinooceanland.com/en-US/About/AboutIndex; "COSCO Int'l to Dispose of All Stake in Sino-Ocean Land," *China Knowledge*, December 17, 2010, http://www.chinaknowledge.com/Newswires/News_Detail.aspx?NewsID=39647; "Morrison & Foerster Advises BOCI and UBS in Sino-Ocean Land Holdings Limited's US$684 Share Placement," Morrison Foerster, January 24, 2011, https://www.mofo.com/resources/press-releases/morrison-foerster-advises-boci-and-ubs-in-sino-ocean-land-holdings-limiteds-us684-share-placement.pdf.

102 Howard J. Dooley, "The Great Leap Outward: China's Maritime Renaissance," *Journal of East Asian Affairs* 26 (Spring/Summer 2012): 53–76, Questia.

103 "Project List: Ocean Plaza (Beijing)," Sino-Ocean Land, n.d., http://www.sinooceanland.com/en-US/BuyHouse/ObjectIndex/?id=455f8e72–8b62–4149-b6ff-9d79699df505.

104 Gurpreet S. Khurana, "China's Maritime Strategy and India: Consonance and Discord," *Maritime Affairs* 7, no. 2 (Winter 2011): 57.

105 "(S) Shipment of Material to Syrian Weapons Entity," Public Library of US Diplomacy, *WikiLeaks*, May 30, 2008, https://wikileaks.org/plusd/cables/08STATE58193_a.html.

106 "Colombia Detains China Cosco Shipping Vessel over Illegal Arms," Reuters, March 3, 2015, http://www.reuters.com/article/us-colombia-china-arms-idUSKBN0LZ29W20150304.

107 Annie-Marie O'Connor and Jeff Leeds, "U.S. Agents Seize Smuggled Arms," *Los Angeles Times*, March 15, 1997, http://articles.latimes.com/1997–03–15/news/mn-38557_1_arms-smuggling.

108 Andrew S. Erickson, *Chinese Naval Shipbuilding: An Ambitious and Uncertain Course* (Annapolis, MD: Naval Institute Press, 2016), 285.

109 Eli Lake, "Kerry's Soft Words Blunt U.S. Hard Power in South China Sea," *Bloomberg*, September 2, 2016, https://www.bloomberg.com/view/articles/2016–09–02/kerry-s-soft-words-blunt-u-s-hard-power-in-south-china-sea.

110 Stuart Grudgings and Lesley Wroughton, "U.S. Gives Tacit Backing to Philippines in China Sea Dispute," Reuters, October 10, 2013, http://www.reuters.com/article/us-asia-summit-idUSBRE9990FD20131010.

111 "Beijing Wants US out of South China Sea Negotiations," *The American Interest*, August 18, 2016, https://www.the-american-interest.com/2016/08/18/beijing-looks-to-keep-us-out-of-scs-negotiations/.

112 "'No Containment' to Be Tested by Time," *Global Times*, February 15, 2014, http://www.globaltimes.cn/content/842572.shtml#.Uv5mK7SAmSo.

113 "Wang Yi Meets with US Secretary of State John Kerry," Embassy of the People's Republic of China in the Republic of Lithuania, January 23, 2014, http://www.chinaembassy.lt/eng/xwdt/t1124337.htm.

114 Bryan Bender, "John Kerry's Brand of Boston Diplomacy Pays Off," *Boston Globe*, November 14, 2014, https://www.bostonglobe.com/news/nation/2014/11/14/secretary-state-john-kerry-brand-boston-diplomacy-pays-off/MMTJZKyos77wBcQWXgdxEL/story.html.

115 Gemini Investments (Holdings) Annual Report 2016, 110–111, http://202.66.146.82/listco/hk/gemini/annual/2016/ar2016.pdf.

116 Josh Rogin, "Chinese Military Sees U.S. as Weak," June 3, 2015, *Bloomberg News*, guest column in *Albuquerque Journal*, https://www.abqjournal.com/593424/chinese-military-sees-us-as-weak.html.

117 Krasnow, "Global Development Firm Purchases Rosemont Realty."

118 "Gemini Investments and Rosemont Realty Form Joint Venture," Rose-

mont Realty, LLC, August 24, 2015, http://www.prnewswire.com/news-releases/gemini-investments-and-rosemont-realty-form-joint-venture-300132117.html.

119 Ibid.

120 Ibid.

CHAPTER 3: NUCLEAR AND OTHER CONSEQUENCES

1 Chao Deng, "Bohai, Harvest and U.S. Investment Firms Expand Target for Outbound Fund," *Wall Street Journal*, July 10, 2014, http://www.wsj.com/articles/bohai-harvest-and-u-s-investment-firms-expand-target-for-outbound-fund-1404956572?tesla=y&mg=reno64-wsj.

2 "Devon Archer—Guiding Diverse International Business Operations," on Devon Archer's personal site, accessed February 22, 2017, http://www.devon-archer.com.

3 "Company Overview of AVIC Automobile Industry Co., Ltd.," *Bloomberg*, accessed February 21, 2017, http://www.bloomberg.com/research/stocks/private/snapshot.asp?privcapId=54884204; "BHR and AVIC Auto Acquire Henniges Automotive," *PR Newswire*, September 15, 2015, http://www.prnewswire.com/news-releases/bhr-and-avic-auto-acquire-henniges-automotive-300143072.html.

4 "History," AVIC website, accessed February 21, 2017, http://www.avic.com/en/aboutus/history/; Robert Wall and Doug Cameron, "Two Chinese Defense Firms Rise in Industry Ranking," *Wall Street Journal*, December 14, 2015, https://www.wsj.com/articles/two-chinese-defense-firms-rise-in-industry-ranking-1450112232.

5 Amrietha Nellan, "AVIC International a Success: How Regulatory Changes to CFIUS Has Limited Political Interference and Empowered Chinese Investors to Obtain a Successful Review," *Hastings Business Law Journal* 9, no. 3 (Spring 2013): 517–538, Nexis.

6 Carl Roper, "PRC Acquisition of US Technology," in Roper, *Trade Secret Theft, Industrial Espionage, and the China Threat* (Boca Raton, FL: CRC Press, 2013), Google Books Reader, 43–55, accessed February 22, 2017; "Military Aviation and Defense," AVIC, accessed February 21, 2017, http://www.avic.com/en/forbusiness/militaryaviationanddefense/index.shtml.

7 "China's Cyber-Theft Jet Fighter," *Wall Street Journal*, November 12, 2014, https://www.wsj.com/articles/chinas-cyber-theft-jet-fighter-1415838777.

8 Arthur Holland Michel, "China's Drones," Center for the Study of the Drone, Bard College, June 22, 2015, http://dronecenter.bard.edu/chinas-drones/.

9 "China's Industry Working with Military," UPI Security Industry April 1, 2011, Ebsco.

10 "BHR and AVIC Auto Acquire Henniges Automotive."

11 "Dual Use Export Licenses," Bureau of Industry and Security, U.S. Department of Commerce, n.d., https://www.bis.doc.gov/index.php/licensing/forms-documents/doc_download/91-cbc-overview.

12 Brad Dawson, "Metzeler, GDX Now Known as Henniges," *Automotive News*, January 28, 2008, http://www.autonews.com/article/20080128/OEM/301289907/metzeler-gdx-now-known-as-henniges; "Homepage," Aerojet Rocketdyne Holdings, Inc., website, accessed February 21, 2017, http://www.aerojetrocketdyne.com/.

13 Robert Woodard, "Commerce Control List Index," Bureau of Industry and Security, U.S. Department of Commerce, September 9, 2011, http://www.bis.doc.gov/index.php/forms-documents/doc_view/13-commerce-control-list-index.

14 "Miller Canfield Advises Chinese Auto Supplier in Largest U.S. Acquisition in History by a Chinese Company," *PR Newswire*, September 10, 2015, http://www.prnewswire.com/news-releases/miller-canfield-advises-chinese-auto-supplier-in-largest-us-acquisition-in-history-by-a-chinese-company-300141284.html.

15 *Report of the Select Committee on U.S. National Security and Military/Commercial Concerns with the People's Republic of China*, H.R. Rep., at 751 (1999).

16 "BHR Newsletter June 2015," BHR Partners Internal Publications, http://en.bhrpe.com/newsitem/276845997.

17 "Composition of CFIUS," U.S. Department of the Treasury, December 1, 2010, https://www.treasury.gov/resource-center/international/foreign-investment/Pages/cfius-members.aspx.

18 Jeff Bennett, "Aviation Industry Corp. of China Completes Buy of Henniges," *Wall Street Journal*, September 9, 2015, https://www.wsj.com/articles/aviation-industry-corp-of-china-completes-buy-of-henniges-1441806079.

19 "Participating as an Anchor Investor in Guangdong Nuclear Power in Hong Kong Listed [translation from Chinese]," BHR Partners website, accessed February 22, 2017, http://www.bhrpe.com/_d276753849.htm.

20 European Commission, Case M. 7850, 1596 Final, October 3, 2016, http://ec.europa.eu/competition/mergers/cases/decisions/m7850_429_3.pdf.

21 "Team Portfolio: Sichuan Sanzhou Special Steel Pipe Co., Ltd.," BHR Partners, accessed February 22, 2017, http://en.bhrpe.comnewsitem/276562348; "Company Overview of Sichuan Sanzhou Special Steel Pipe Co., Ltd.," *Bloomberg*, accessed February 22, 2017, http://www.bloomberg.com/research/stocks/private/snapshot.asp?privcapid=134193940.

22 Jamie Satterfield, "Records: Former TVA Manager Admits Chinese Government Paid Him for Nuclear Secrets," *Knoxville News Sentinel*, April 29, 2016, http://archive.knoxnews.com/news/crime-courts/chinese-engineer-accused-in-nuclear-espionage-case-returned-to-knoxville-31a0ea9f-8ff8-2d97-e053-010—377609161.html/.

23 Ibid.; "Distance from Allen Ho's House to Joe Biden's House," accessed on

Google Maps, February 22, 2017, https://www.google.com/maps/dir/1209+Barley+Mill+Rd/302+N+Ashview+Ln/@39.7753586,-75.6162048,14z/data=!3m1!4b1!4m14!4m13!1m5!1m1!1s0x89c6fdd21aa17c3d:0xbc95e1f8ea405e1e!2m2!1d-75.609975!2d39.770358!1m5!1m1!1s0x89c6fc45fd0ffa8f:0xf3e4564fcae116a8!2m2!1d-75.591575!2d39.783072!3e0.

24 "U.S. Nuclear Engineer, China General Nuclear Power Company and Energy Technology International Indicted in Nuclear Power Conspiracy Against the United States," U.S. Department of Justice, April 14, 2016, https://www.justice.gov/opa/pr/us-nuclear-engineer-china-general-nuclear-power-company-and-energy-technology-international.

25 *2016 Report to Congress of the U.S.-China Economic and Security Review Commission*, 114th Cong., 2nd Sess. (November 2016), https://www.uscc.gov/sites/default/files/annual_reports/2016%20Annual%20Report%20to%20Congress.pdf.

26 Bloomberg News, "China Climate Pledge Needs 1,000 Nuclear Plant Effort," *Bloomberg*, November 20, 2014, https://www.bloomberg.com/news/articles/2014–11–21/latest-china-revolution-seeks-great-leap-for-clean-energy.

27 "U.S. Charges Five Chinese Military Hackers for Cyber Espionage Against U.S. Corporations and a Labor Organization for Commercial Advantage," U.S. Department of Justice, May 19, 2014, https://www.justice.gov/usao-wdpa/pr/us-charges-five-chinese-military-hackers-cyber-espionage-against-us-corporations-and.

28 "China Climate Pledge Needs 1,000 Nuclear Plant Effort."

29 European Trade Commission, "Annex 1: List of Dual-Use Items and Technology," referred to in Article 3, Regulation (EC) No. 1334/2000, http://trade.ec.europa.eu/doclib/docs/2008/september/tradoc_140595.pdf.

30 Steven Mufson, "Obama's Quiet Nuclear Deal with China Raises Proliferation Concerns," *Washington Post*, May 10, 2015, https://www.washingtonpost.com/business/economy/obamas-quiet-nuclear-deal-with-china-raises-proliferation-concerns/2015/05/10/549e18de-ece3–11e4–8666-a1d756d0218e_story.html?utm_term=.f282e1832970.

31 "U.S. Nuclear Engineer Pleads Guilty to Violating the Atomic Energy Act," U.S. Department of Justice, January 6, 2017, https://www.justice.gov/usao-edtn/pr/us-nuclear-engineer-pleads-guilty-violating-atomic-energy-act.

32 "U.S. Nuclear Engineer, China General Nuclear Power Company and Energy Technology International Indicted in Nuclear Power Conspiracy Against the United States."

33 Dr. Allen Ho, interview by FBI, April 14, 2016, Case 3:16-cr-00046-TAV-HBG, Document 52–1, transcript, filed September 12, 2016.

34 Ibid.

35 Ralph Sawyer, *The Tao of Spycraft: Intelligence Theory and Practice in Traditional China* (Boulder, CO: Westview Press, 1998), xiii. Note: China has been aggressively stealing U.S. military technology secrets for years. A

"damage assessment" following Chinese espionage in the 1990s headed by Admiral David Jeremiah concluded that classified information obtained by the Chinese government "probably accelerated its program to develop future nuclear weapons." And their method of collection is different from other countries', especially when it comes to so-called S&T (science and technology) espionage. Chinese officials don't rely on Chinese intelligence services to obtain the information, but on companies, institutes, and other entities to collect it themselves and then share it with the government. The Chinese government has a history of trying to steal secret information on American nuclear reactor components. Less than a year earlier, in May 2014, five Chinese military hackers were indicted for stealing internal documentation from several Western nuclear companies, including Westinghouse. Between 2006 and 2014 the hackers tried to access designs for "pipes, pipe supports, and pipe routing within the AP1000 plant buildings." The fact that Chinese government hackers and CGN were both targeting the same reactor for secrets was no coincidence. "What you hear again and again from foreign company executives working in China is that the Chinese are absolutely determined to have as much technology transferred from foreign entities as they can," said Mark Hibbs, a senior associate at the Carnegie Endowment for International Peace.

36 Graham Park, *Introducing Natural Resources* (Edinburgh: Dunedin, 2016), 29, Questia.

37 Michael Montgomery, "Molybdenum's Future Demand Strong from Nuclear Power Application," *Molybdenum Investing News*, December 15, 2011, http://investingnews.com/daily/resource-investing/industrial-metals-investing/molybdenum-investing/molybdenums-future-demand-strong-from-nuclear-power-application/.

38 *2015 Annual Report*, China Molybdenum Co., Ltd., http://www.chinamoly.com/06invest/DOC/E_03993_AR009_0426.pdf, 42.

39 "Company News: Resonant Red Songs Resounded Through Luomu," China Molybdenum Co., Ltd., July 18, 2010, http://www.chinamoly.com/en/05news/detail_new_100.htm.

40 "Luoyang Mining PPN Creditors Demand Early Redemption," *Asiamoney*, February 2014, ProQuest; Hal Quinn, "U.S. Manufacturers Need Access to Metals," *National Defense*, July 2014, Questia.

41 Sudeep Reddy, "Trade Battle Gives Obama a Venue on China," *Wall Street Journal Online*, March 13, 2012, Factiva.

42 "When China Bends the Rules," *International New York Times*, March 28, 2014, Questia; Umair Ghori, "An Epic Mess: 'Exhaustible Natural Resources' and the Future of Export Restraints After the China–Rare Earths Decision," *Melbourne Journal of International Law* 16, no. 2 (December 2015): 1–34, Questia.

43 Anet Josline Pinto and Denny Thomas, "Freeport to Sell Prized Tenke Copper Mine to China Moly for $2.65 Billion," Reuters, May 9, 2016,

http://www.reuters.com/article/us-freeport-mcmoran-tenke-cmoc-idUSKCN0Y015U.

44 Sunny Freeman, "Lundin Mining Sells Stake in Giant Congolese Copper Mine for $1.5 Billion," *Financial Post*, November 15, 2016, http://business.financialpost.com/news/mining/lundin-mining-sells-stake-in-congo-mine-owner-to-chinese-firm-for-us1–14-billion.

45 "China Moly to Help BHR Acquire Stake in Congo's Tenke Copper Mine," Reuters, January 22, 2017, http://www.reuters.com/article/us-congo-mining-idUSKBN1560OP.

46 "China Plays Long Game on Cobalt and Electric Batteries," *Financial Times*, May 26, 2016, https://www.ft.com/content/054bbb3a-1e8b-11e6-a7bc-ee846770ec15.

CHAPTER 4: BIDENS IN UKRAINE

1 James Stafford, "Bribery, Back Room Dealing and Bullying in Ukraine: The Origins of Burisma," *OilPrice.com*, October 15, 2015, accessed March 1, 2017, Wayback Machine; Ilya Timtchenko, "British Court Unfreezes Accounts of Yanukovych-era Ecology Minister Zlochevsky," *Kyiv Post*, January 23, 2015, https://www.kyivpost.com/article/content/ukraine-politics/british-court-unfreezes-accounts-of-yanukovych-era-ecology-minister-zlochevsky-378238.html; Victoria Parker, "Shale Gas and the War in East Ukraine," *Axis of Logic*, July 12, 2014, http://axisoflogic.com/artman/publish/Article_67005.shtml.

2 Roman Goncharenko, "Who Are Hunter Biden's Ukrainian Bosses?" *DW* (Bonn), May 16, 2014, http://www.dw.com/en/who-are-hunter-bidens-ukrainian-bosses/a-17642254.

3 Katya Gorchinskaya and Serhiy Andrushko, "Former Ukrainian Official on the Lam in Alligator Shoes?" *Radio Free Europe*, July 31, 2015, http://www.rferl.org/a/ukraine-zlocevskiy-zlocci-shoe-store/27163483.html.

4 Dmytro Hnap and Anna Babinets, "Kings of Ukrainian Gas," *Anti-Corruption Action Centre*, August 26, 2012, https://antac.org.ua/2012/08/26/kings-of-ukrainian-gas/.

5 Shaun Walker, "Ukraine Elections: High Stakes and Dirty Tricks in Hotly Contested Vote," *Guardian* (Manchester), October 21, 2015, https://www.theguardian.com/world/2015/oct/21/ukraine-elections-dirty-tricks.

6 "Kolomoisky Seen as in Charge of 23-Member Vidrodzhennya Bloc," *Kyiv Post*, October 28, 2016, https://www.pressreader.com/ukraine/kyiv-post/20161028/281689729360531; Eric Ellis, "Ukraine: The Perils of PrivatBank," *Euro Money*, April 29, 2016, http://www.euromoney.com/Article/3550245/Ukraine-The-perils-of-PrivatBank.html; Joshua Yaffa, "Reforming Ukraine After the Revolutions," *New Yorker*, September 5, 2016, http://www.newyorker.com/magazine/2016/09/05/reforming-ukraine-after-maidan.

7 “Ukraine’s Tycoon and Governor Kolomoisky Confesses to Holding 3 Passports,” *Tass*, October 3, 2014, http://tass.com/world/752691.

8 Jim Armitage, “Raiders from the East: The Oligarchs Who Won Their Case but Took a Battering,” *Independent* (London), September 10, 2013, http://www.independent.co.uk/news/business/analysis-and-features/raiders-from-the-east-the-oligarchs-who-won-their-case-but-took-a-battering-8807681.html.

9 David Barrett, “Ukrainian Oligarchs Clash in Court over $2bn Business Deal Amid Claims of Murder and Bribery,” *Telegraph*, December 4, 2015, http://www.telegraph.co.uk/news/uknews/law-and-order/12034304/Ukrainian-oligarchs-clash-in-court-over-2bn-business-deal-amid-claims-of-murder-and-bribery.html.

10 Melik Kaylan, “An Injection of Rule of Law for Ukrainian Business? Oligarch’s Lawsuit Could Help Improve the Culture of Business Dealings in the Post Soviet Space,” *Forbes*, July 15, 2013, http://www.forbes.com/sites/melikkaylan/2013/07/15/an-injection-of-rule-of-law-for-ukrainian-business-oligarchs-lawsuit-could-help-improve-the-culture-of-business-dealings-in-the-post-soviet-space/#2843445527b2.

11 James Nadeau, “Ukraine’s Real Problem: Crony Capitalism,” *The Hill*, January 15, 2014, http://thehill.com/blogs/congress-blog/foreign-policy/195549-ukraines-real-problem-crony-capitalism.

12 Andrew Cockburn, “Undelivered Goods: How $1.8 Billion in Aid to Ukraine Was Funneled to the Outposts of the International Finance Galaxy,” *Harper’s Magazine*, August 13, 2015, http://harpers.org/blog/2015/08/undelivered-goods/.

13 Ibid.

14 Scott Neuman and L. Carol Ritchie, “Ukrainian President Voted Out; Opposition Leader Freed,” *NPR.org*, February 22, 2014, http://www.npr.org/sections/thetwo-way/2014/02/22/281083380/unkrainian-protesters-uneasy-president-reportedly-leaves-kiev; Shaun Walker, “Ousted Ukrainian Leader Viktor Yanukovych Reported to Be in Russia,” *Guardian* (Manchester), February 27, 2014, https://www.theguardian.com/world/2014/feb/27/viktor-yanukovych-russia-ukrainian-president-moscow.

15 David Von Drehle, et al., “What Putin Wants,” *Time,* 183, no. 10 (March 17, 2014): 22, EBSCOhost; Nolan Peterson, “Ukraine Arms Deal Deals Blow to Putin Backed Rebels,” *Newsweek*, August 7, 2017, http://newsweek.com/ukraine-arms-deal-deals-blow-putin-backed-rebels-647276.

16 Martin Nunn and Martin Foley, “Why Does Putin Want Crimea Anyway?” *EU Observer,* March 17, 2014, https://euobserver.com/opinion/123496.

17 “‘Cyber Berkut’ Hackers Target Major Ukrainian Bank,” *Moscow Times*, July 4, 2014, http://old.themoscowtimes.com/business/article/cyber-berkut-hackers-target-major-ukrainian-bank/502992.html.

18 Javier Jarrín, “International Response to Annexation of Crimea,” *Euro-*

maidan Press, March 24, 2014, http://euromaidanpress.com/2014/03/24/international-response-to-annexation-of-crimea/#arvlbdata.

19 Michael R. Gordon, "Kerry Takes Offer of Aid to Ukraine and Pushes Back at Russian Claims," *New York Times*, March 4, 2014, https://www.nytimes.com/2014/03/05/world/europe/secretary-of-state-john-kerry-arriving-in-kiev-offers-1-billion-in-loan-guarantees-to-ukraine.html.

20 Ibid.

21 Michael Scherer, "Ukrainian Employer of Joe Biden's Son Hires a D.C. Lobbyist," *Time*, July 8, 2014, http://time.com/2964493/ukraine-joe-biden-son-hunter-burisma/.

22 Dan De Luce and Reid Standish, "What Will Ukraine Do Without Uncle Joe?" *Foreign Policy*, October 30, 2016.

23 Leonid Bershidsky, "Biden Tells Ukraine to . . . Do Nothing," *Bloomberg*, December 8, 2015, https://www.bloomberg.com/view/articles/2015–12–08/biden-tells-ukraine-to-do-nothing.

24 Steve Clemons, "One Last Trip with Joe Biden," *Atlantic*, January 19, 2017, https://www.theatlantic.com/international/archive/2017/01/joe-biden-foreign-policy/513758/.

25 "White House Visitor Logs," InsideGov, http://white-house-logs.insidegov.com.

26 Stafford, "Bribery, Back Room Dealing and Bullying in Ukraine."

27 "United States Supports Ukraine's Energy Independence," Burisma Group website, September 22, 2014, http://burisma.com/en/media/united-states-supports-ukraine-s-energy-independence/.

28 "Fact Sheet: U.S. Crisis Support Package for Ukraine," The White House Office of the Press Secretary, April 21, 2014, https://obamawhitehouse.archives.gov/the-press-office/2014/04/21/fact-sheet-us-crisis-support-package-ukraine; JC Finley, "Vice President Biden arrives in Kiev," *United Press International,* April 21, 2014, http://www.upi.com/Top_News/World-News/2014/04/21/Vice-President-Biden-arrives-in-Kiev/4651398098183/.

29 Kate Stanton, "Hunter Biden Joins Board of Ukraine Gas Company," UPI, May 14, 2014, http://www.upi.com/Top_News/US/2014/05/14/Hunter-Biden-joins-board-of-Ukraine-gas-company/1791400108748/.

30 Ibid.

31 Scherer, "Ukrainian Employer of Joe Biden's Son Hires a D.C. Lobbyist."

32 "No Issue with Biden's Son, Ukraine Gas Firm: White House," *China Daily* (Beijing), May 15, 2014, http://www.chinadaily.com.cn/world///////2014-05/15/content_17509096.htm; "Biden's Son Gets Energy Job," *China Daily* (Beijing), May 15, 2014, http://www.chinadaily.com.cn/world/2014-05/19/content_17517458.htm.

33 Scherer, "Ukrainian Employer of Joe Biden's Son Hires a D.C. Lobbyist."

34 Ibid.

35 "Burisma Holdings Attends US Congress Meeting with President Poroshenko and Meets with US Senators on Energy and Environment," Burisma

Group website, April 1, 2016, http://burisma.com/en/news/burisma-holdings-attends-us-congress-meeting-with-president-poroshenko-and-meets-with-us-senators-on-energy-and-environment/.

36 Stafford, "Bribery, Back Room Dealing and Bullying in Ukraine."

37 "British Court Unblocks Accounts of Ex-Minister Zlochevsky with $23 Million, Says Burisma," *Ukraine General Newswire*, January 23, 2015, http://en.interfax.com.ua/news/general/246364.html. James Risen, "Joe Biden, His Son and the Case Against a Ukrainian Oligarch," New York Times, December 8, 2015, https://www.nytimes.com/2015/12/09/world/europe/corruption-ukraine-joe-biden-son-hunter-biden-ties.html.

38 Klaus Schwab and Xavier Sala-i-Martín, *The Global Competitiveness Report: 2014–2015*, Insight Report, World Economic Forum, 2014, http://www3.weforum.org/docs/WEF_GlobalCompetitivenessReport_2014–15.pdf.

39 Josh Cohen, "Corruption in Ukraine Is So Bad, a Nigerian Prince Would Be Embarrassed," Reuters, December 30, 2015, http://blogs.reuters.com/great-debate/2015/12/30/corruption-in-ukraine-is-so-bad-a-nigerian-prince-would-be-embarrassed-2/.

40 Stafford, "Bribery, Back Room Dealing and Bullying in Ukraine."

41 Ibid.

42 "Burisma Group Together with the Prince Albert II of Monaco Foundation Will Be Holding a European Forum on Energy Security for the Future: New Sources, Responsibility, Sustainability," Burisma Group website, press release, April 20, 2016, http://burisma.com/en/news/burisma-group-together-with-the-prince-albert-ii-of-monaco-foundation-will-be-holding-a-european-forum-on-energy-security-for-the-future-new-sources-responsibility-sustainability/.

43 Oliver Boyd-Barrett, *Western Mainstream Media and the Ukraine Crisis: A Study in Conflict Propaganda* (New York: Routledge, 2017), 56.

44 Cockburn, "Undelivered Goods."

45 Vice President Joe Biden and Ukrainian President Petro Poroshenko, "Remarks by the Vice President Joe Biden and Ukrainian President Petro Poroshenko" (speech, The Bankova, Kyiv, Ukraine, December 7, 2015), The White House Office of the Vice President, https://obamawhitehouse.archives.gov/the-press-office/2015/12/07/remarks-vice-president-joe-biden-and-ukrainian-president-petro; Seung Min Kim, "Kerry: Include IMF in Ukraine Bill," *Politico*, March 13, 2014, https://www.politico.com/story/2014/03/john-kerry-include-imf-changes-in-ukraine-package-104629.

46 Ibid.

47 Ibid.

48 Ibid.

49 Graham Stack, "Privat Investigations: PrivatBank Lending Practices

Threaten Ukraine's Financial Stability," *bne IntelliNews*, December 19, 2016, http://www.intellinews.com/privat-investigations-privatbank-lending-practices-threaten-ukraine-s-financial-stability-112477/.

50 Natalia Zinets and Pavel Polityuk, "Ukraine's Largest Bank Rescued by State, Poroshenko Urges Depositors to Stay Calm," Reuters, December 19, 2016, http://www.reuters.com/article/us-ukraine-crisis-privatbank-idUSKBN1480MY.

51 American Center for a European Ukraine, "GPO Seizes Property of Ex-Minister Zlochevsky," February 5, 2016, http://www.europeanukraine.org/home/2016/02/gpo-seizes-property-of-ex-minister-zlochevsky/.

52 Stafford, "Bribery, Back Room Dealing and Bullying in Ukraine."

53 "The Court Repeatedly Seized Wells the Company Zlochevsky," *NEWS.ru*, accessed March 1, 2017, http://en.few-news.ru/the-court-repeatedly-seized-wells-the-company-zlochevsky.html.

54 "In Borispol the Plane Landed, Vice-President of the USA Joe Biden," *Ukrop News 24*, photograph, January 16, 2017, http://ukropnews24.com/in-borispol-the-plane-landed-vice-president-of-the-usa-joe-biden/.

55 Pavel Polityuk and Alessandra Prentice, "U.S. Vice President Biden to Make Swansong Visit to Ukraine," Reuters, January 12, 2017, http://www.reuters.com/article/us-ukraine-crisis-biden-idUSKBN14W0QT.

56 Alyona Zhuk, "Biden to Visit Ukraine Again on January 15, 2016," *Kyiv Post*, January 13, 2017.

57 Burisma Group, "All Cases Closed Against Burisma Group and Its President Nikolay Zlochevskyi in Ukraine. The Company Cooperated with Law Enforcement Agencies and Paid in Full All Outstanding Fees," *Kyiv Post*, January 12, 2017, https://www.kyivpost.com/business-wire/cases-closed-burisma-group-president-nikolay-zlochevskyi-ukraine-company-cooperated-law-enforcement-agencies-paid-full-outstanding-fees.html.

58 "On Final Ukraine Trip, Biden Urges Trump Administration to Keep Russia Sanctions," *Guardian* (Manchester), January 16, 2017, https://www.theguardian.com/us-news/2017/jan/16/joe-biden-ukraine-visit-russia-sanctions.

59 American Indian Humanitarian Foundation, "Pine Ridge Statistics," accessed March 6, 2017, http://www.4aihf.org/id40.html.

60 Regina F. Graham, "The 'Porn King' of LA 'Ripped Off Impoverished Sioux Tribe in $60 Million Scam,'" *Daily Mail*, May 11, 2016, http://www.dailymail.co.uk/news/article-3585626/Man-dubbed-Porns-New-King-charged-tribal-fraud.html.

61 Securities and Exchange Commission v. Devon D. Archer, Bevan T. Cooney, Hugh Dunkerley, Jason W. Galanis, John P. Galanis, Gary T. Hirst and Michelle A. Morton, May 11, 2016, p. 29, section 97, https://www.sec.gov/litigation/complaints/2016/comp-pr2016–85.pdf.

62 Sealed Complaint: USA v. Jason Galanis, Gary Hirst, John Galanis, Hugh Dunkerley, Michelle Morton, Devon Archer, and Bevan Cooney, p. 4, https://www.justice.gov/usao-sdny/file/850241/download.

63 Sealed Complaint: Securities and Exchange Commission v. Devon D. Archer, Bevan T. Cooney, Hugh Dunkerley, Jason W. Galanis, John P. Galanis, Gary T. Hirst and Michelle A. Morton, May 11, 2016, p. 11, section 29.

64 Bernard Condon, "The Long Reach of John Peter Galanis," *Forbes*, September 18, 2000, https://www.forbes.com/forbes/2000/0918/6608186a.html.

65 Seth Lubove, "Porn's New King," *Forbes*, March 22, 2004, https://www.forbes.com/2004/03/22/cz_sl_0322galanis.html.

66 Condon, "The Long Reach of John Peter Galanis."

67 Sealed Complaint: USA v. Jason Galanis, Gary Hirst, John Galanis, Hugh Dunkerley, Michelle Morton, Devon Archer, and Bevan Cooney, p. 13.

68 Ibid.

69 Jason Galanis, e-mail message to Michelle Morton, June 3, 2014, "Introducing COR Capital Summary" (unpublished manuscript), Microsoft Word file, accessed on February 24, 2017, http://www.bancexposed.com/wp-content/uploads/2016/10/Galanis-COR-Email-1.pdf.

70 Sealed Complaint: Securities and Exchange Commission v. Devon D. Archer, Bevan T. Cooney, Hugh Dunkerley, Jason W. Galanis, John P. Galanis, Gary T. Hirst and Michelle A. Morton, p. 29, section 60.

71 Ibid., p. 29, sections 96–97.

72 Sealed Complaint: USA v. Jason Galanis, Gary Hirst, John Galanis, Hugh Dunkerley, Michelle Morton, Devon Archer, and Bevan Cooney, p. 14, section 33.

73 Ibid., p. 30, section 47.

74 Ibid., p. 31, section 47.

75 "SEC Charges Individual Who Headed Fake Investment Manager Used in Tribal Bonds Scheme," Securities and Exchange Commission, November 16, 2016, https://www.sec.gov/litigation/litreleases/2016/lr23689.htm.

76 Sealed Complaint: Securities and Exchange Commission v. Devon D. Archer, Bevan T. Cooney, Hugh Dunkerley, Jason W. Galanis, John P. Galanis, Gary T. Hirst and Michelle A. Morton, p. 29.

77 Aloha Startups, "HSDC and mbloom Close $10M Startup Investment Fund," January 22, 2014, http://www.alohastartups.com/2014/01/22/hsdc-and-mbloom-close-10m-startup-investment-fund/.

78 Jason Ubay, "Stock Price for Maui-Based Software Firm Code Rebel Soars After IPO," *Pacific Business News*, May 21, 2015, http://www.bizjournals.com/pacific/news/2015/05/20/stock-price-for-maui-based-software-firm-code.html.

79 Sealed Complaint: Securities and Exchange Commission v. Devon D. Archer, Bevan T. Cooney, Hugh Dunkerley, Jason W. Galanis, John P. Galanis, Gary T. Hirst and Michelle A. Morton, pp. 31–32, sections 104–108.

80 "SEC Charges Father, Son, Others in Tribal Bonds Scheme," U.S. Securities and Exchange Commission, May 11, 2016, https://www.sec.gov/litigation/litreleases/2016/lr23535.htm. Sealed Complaint: USA v. Jason Galanis, Gary Hirst, John Galanis, Hugh Dunkerley, Michelle Morton, Devon Archer, and Bevan Cooney, p. 1.

81 Sealed Complaint: The Michelin Retirement Plan and Investment Committee of the Michelin Retirement Plan v. Dilworth Paxson, et al., https://www.unitedstatescourts.org/federal/scd/232086/1–0.html.

82 "Bohai Huamei (Shanghai) Equity Investments," Investment Fund Management Co., Ltd., May 20, 2016, https://translate.google.com/translate?hl=en&sl=zh-CN&u=http://company.xizhi.com/GS5707a9631f98cc0d678b461e/&prev=search.

83 "Monitoring: Burisma Holdings Board Director Archer Arrested," Ukrainian News Agency (Kiev), May 13, 2016.

84 Associated Press, "Porn's New King Sentenced to over 14 Years in Prison," *New York Post*, August 11, 2017, https://nypost.com/2017/08/11/porns-new-king-sentenced-to-over-14-years-in-prison/

CHAPTER 5: McCONNELL AND CHAO: FROM CHINA WITH PROFITS

1 Matt Egan, "Trump Transportation Pick Elaine Chao Made $1.2 Million from Wells Fargo," *CNN Money*, December 7, 2016, http://money.cnn.com/2016/12/07/investing/elaine-chao-trump-wells-fargo-transportation/; "Early Career 1983–89," Secretary Elaine L. Chao website, accessed April 4, 2017, http://www.elainelchao.com/early-career/.

2 Josie Albertson-Grove and Masako Hirsch, "Family's Shipping Company Could Pose Problems for Trump's Transportation Pick," *ProPublica*, December 12, 2016, https://www.propublica.org/article/familys-shipping-company-could-pose-problems-for-trumps-transportation-pick.

3 "About Dr. James S. C. Chao," Foremost Foundation website, accessed April 4, 2017, http://www.theforemostfoundation.org/dr-james-sc-chao.

4 Elaine L. Chao, "Ordinary Yet Extraordinary—Ruth Mulan Chu Chao's Story," *Asian Fortune News*, updated October 1, 2007, accessed April 4, 2017, http://www.asianfortunenews.com/site/article_1007.php?article_id=59.

5 "About Dr. James S. C. Chao."

6 Brendan Morrow, "Elaine Chao's Sisters: 5 Fast Facts You Need to Know" *Heavy.com*, November 29, 2016, http://heavy.com/news/2016/11/elaine-chao-sisters-family-siblings-jeanette-may-christine-grace-angela-james-ruth-family-history-harvard/; "About Mrs. Ruth Mulan Chu Chao," Foremost Foundation website, accessed April 4, 2017, http://www.theforemostfoundation.org/mrs-ruth-mulan-chu-chao.

7 "Early Career," Secretary Elaine L. Chao website.

8 Alec MacGillis, "Chapter Two: No Money Down" in MacGillis, *The Cynic: The Political Education of Mitch McConnell* (New York: Simon & Schuster Paperbacks, 2014), 58.

9 Lee Siew Hua, "Stories My Father Told Me," *Straits Times*, August 12, 2001, Nexis.

10 Mae Cheng, "Immigrant's Success Story; Labor Nominee Has Queens Roots," *Newsday*, January 14, 2001, Nexis.

11 John B. Judis, "Sullied Heritage (Cover Story)," New Republic 224, no. 17 (2001): 19–25, Academic Search Ultimate, EBSCOhost, accessed March 27, 2017.

12 Ibid.

13 "Chinese Vice Premier Meets U.S. Guests," Xinhua News Agency, December 31, 1993, Nexis.

14 Nicholas D. Kristof, "Crackdown in Beijing; Troops Attack and Crush Beijing Protest; Thousands Fight Back, Scores Are Killed," *New York Times*, June 4, 1989, http://www.nytimes.com/1989/06/04/world/crackdown-beijing-troops-attack-crush-beijing-protest-thousands-fight-back.html?ref=tiananmensquare.

15 "Tiananmen Square Fast Facts," *CNN.com*, June 3, 2016, http://www.cnn.com/2013/09/15/world/asia/tiananmen-square-fast-facts/.

16 Judis, "Sullied Heritage"; "Chinese Vice Premier Meets U.S. Guests," Xinhua News Agency, December 31, 1993, Nexis.

17 "Jiang Zemin Meets U.S. Guests," Xinhua News Agency, December 30, 1993; "Chinese Vice Premier," Xinhua News Agency.

18 Judis, "Sullied Heritage," 19–25.

19 Ibid.

20 Ibid.

21 Ibid.

22 Ibid.

23 "Mitch McConnell on China-Voting Record," Political Guide website, accessed April 4, 2017, http://www.thepoliticalguide.com/Profiles/Senate/Kentucky/Mitch_McConnell/Views/China/.

24 Currency Exchange Rate Oversight Reform Act of 2011, S.1619, 112th Cong. (2011–2012), https://www.congress.gov/bill/112th-congress/senate-bill/1619.

25 Pete Kasperowicz, "Senate Puts Pressure on House, China, by Passing Currency Bill," *The Hill*, October 12, 2011, http://thehill.com/blogs/floor-action/senate/186841-senate-pressures-china-house-by-passing-currency-bill.

26 Felicia Sonmez, "Senate Makes Unprecedented Rules Change amid Late-Night Debate over Jobs, Procedure," *Washington Post*, October 7, 2011, https://www.washingtonpost.com/blogs/2chambers/post/senate-makes-unprecedented-rules-change-amid-late-night-debate-over-jobs-procedure/2011/10/06/gIQA91BpRL_blog.html?utm_term=.bba9d1f6f67d; Don

Wolfensberger, "Harry Reid's Nuclear Test Benefits President, for Now," *Roll Call*, October 24, 2011, http://www.rollcall.com/news/harry_reid_nuclear_test_benefits_president_for_now-209719-1.html.

27 Alexander Bolton, "Reid Triggers 'Nuclear Option' to Change Senate Rules, End Repeat Filibusters," *The Hill*, October 7, 2011, http://thehill.com/homenews/senate/186133-reid-triggers-nuclear-option-to-change-senate-rules-and-prohibit-post-cloture-filibusters.

28 United States Senate, S.1619, "U.S. Senate Roll Call Votes 112th Congress—1st Session," October 11, 2011, https://www.senate.gov/legislative/LIS/roll_call_lists/roll_call_vote_cfm.cfm?congress=112&session=1&vote=00159/; Currency Exchange Rate Oversight Reform Act of 2011, Actions Overview, S.1619, 112th Cong. (2011–2012), https://www.congress.gov/bill/112th-congress/senate-bill/1619/all-actions.

29 John Cheves, "Wedded to Free Trade in China," *Lexington Herald-Leader*, (October 20, 2006). Nexis.

30 *Congressional Record-Senate*, May 9, 1995, S6322, https://www.congress.gov/crec/1995/05/09/CREC-1995-05-09-pt1-PgS6322.pdf; Leon Hadar, "U.S. Sanctions against Burma: A Failure on All Fronts," *Cato Institute, Trade Policy Analysis* no.1 March 26, 1998, https://www.cato.org/publications/trade-policy-analysis/us-sanctions-against-burma-failure-all-fronts; Bernie Becker and Vicki Needham, "McConnell Warns Obama Against Carveout in Trade Deal," *The Hill*, July 31, 2015, http://thehill.com./policy/finance/249913-mcconnell-warns-obama-against-tobacco-carveout-in-trade-deal.

31 Anne Applebaum, "'Show of Power,' Indeed," *Washington Post*, August 26, 2008, http://www.washingtonpost.com/wp-dyn/content/article/2008/08/25/AR2008082502333.html.

32 "U.S. Department of Labor Supports USTR Decision on Section 301 Petition Regarding China," *PR Newswire*, July 21, 2006, Nexis.

33 Li Yi, "Elaine Chao Answers Questions from PKU Students," Peking University Newsletter no. 23 (Summer 2013): 17, http://www.doc88.com/p-9435291941544.html.

34 "The Real Elaine Chao," *World Net Daily*, July 16, 2001, http://www.wnd.com/2001/07/10009/.

35 Jeffrey Reeves, *Chinese Foreign Relations with Weak Peripheral States: Asymmetrical Economic Power and Insecurity* (New York: Routledge, 2016), 141.

36 Irene Ang, "Foremost Maritime Orders Newscastlemax Duo at Nacks," *TradeWinds*, November 22, 2013.

37 Irene Ang, "Foremost Orders Pair of Capesizes at Qingdao Beihai," *TradeWinds*, April 11, 2014, http://www.tradewindsnews.com/weekly/335839/foremost-orders-pair-of-capesizes-at-qingdao-beihai; Irene Ang, "Foremost Returns to Beihai for Another Four Capesizes," *TradeWinds*, July 25, 2014, http://www.tradewindsnews.com/weekly/341649/Foremost-returns-to-Beihai-for-another-four-capesizes; Trond Lillestolen, "Fore-

most Maritime Sells Bulker for $8m to $8.2m," *TradeWinds*, September 4, 2015, http://www.tradewindsnews.com/weekly/367527/foremost-maritime-sells-bulker-for-usd-8m-to-usd-82m.

38 "Chinese President Hu Met with Foremost and CSSC Leader," China State Shipbuilding Corporation website, September 25, 2010, http://www.cssc.net.cn/en/component_news/news_detail.php?id=7021.

39 Irene Ang, "Foremost in Shanghai for Two Bulkers," *TradeWinds*, October 6, 2011.

40 Irene Ang, "Foremost on Bulker List for China Capes," *TradeWinds*, September 5, 2002.

41 Irene Ang, "Foremost Back in Shanghai for Capesizes," *TradeWinds*, August 31, 2006.

42 Ang, "Foremost in Shanghai for Two Bulkers."

43 Trond Lillestolen, "Foremost Bags Firm $10.8 Million for Oldest Bunker," *TradeWinds*, February 16, 2017.

44 Andy Pierce, "Foremost Capesize Deal Shows Price Rise," *TradeWinds*, May 16, 2016.

45 Trond Lillestolen, "Capesize Owners Grit Teeth as Rates Reach Rock-Bottom and Beyond," *TradeWinds*, May 19, 2011.

46 Lillestolen, "Foremost Maritime Sells Bulker for $8m to $8.2m."

47 Ang, "Foremost Returns to Beihai for Another Four Capesizes."

48 Albertson-Grove and Hirsch, "Family's Shipping Company Could Pose Problems."

49 For Cargill see Eric Martin, "Capesize Climb," *TradeWinds*, June 4, 2008; for Rizhao Steel see Jim Mulrenan, "Deft Dreyfus," *TradeWinds*, June 3, 2008; for Wuhan see Aaron Kelley, "Foremost Ties Up Two," *TradeWinds*, October 25, 2010.

50 "Angela Chao," Chao Family Foundation's website, accessed April 4, 2017, http://www.chaofamilyfoundations.com/angela-chao/.

51 China State Shipbuilding Corporation, Shanghai stock exchange no. 600150, *Annual Report*, 2008, 10, 11; China State Shipbuilding Corporation, Shanghai stock exchange no. 600150, *Annual Report*, 2010, 12.

52 Gabe Collins and Eric Anderson, "Resources for China's State Shipbuilders: Now Including Global Capital Markets," in *Chinese Naval Shipbuilding: An Ambitious and Uncertain Course,* ed. Andrew S. Erickson (Annapolis, MD: Naval Institute Press, 2016), 63–65; China State Shipbuilding Corporation, Shanghai stock exchange no. 600150, *Annual Report*, 2008; China State Shipbuilding Corporation, Shanghai Stock Exchange no. 600150, *Annual Report*, 2010.

53 Daniel Alderman and Rush Doshi, "Civil-Military Integration Potential in Chinese Shipbuilding," in Erikson, ed., *Chinese Naval Shipbuilding: An Ambitious and Uncertain Course*, ed. Andrew S. Erickson (Annapolis, MD: Naval Institute Press, 2016), 146; Collins and Anderson, "Resources for China's State Shipbuilders," in Erickson, ed., *Chinese Naval Shipbuilding: An Ambitious and Uncertain Course*, 62.

54 "Groups Profile," China State Shipbuilding Corporation website, accessed April 4, 2017, http://www.cssc.net.cn/en/component_general_situation/.

55 Alderman and Doshi, "Civil-Military Integration Potential in Chinese Shipbuilding," 149. Note: The Chaos' background is, of course, in commercial shipping, not military vessels. But in China, the two—civilian and military—are deeply intertwined. The federal government's U.S.-China Economic and Security Review Commission noted that "CSSC's extensive experience in the commercial shipping sector helps to integrate the civil and military arms of the enterprise." The Chinese government views civilian and military shipbuilding as one and the same. As Daniel Alderman of Defense Group Inc. and Rush Doshi of Harvard put it in their academic paper on the company, the Chinese government "views civilian shipbuilding as essential to and intertwined with military shipbuilding." For the Chinese government, shipbuilding is not just a commercial business that creates jobs and wealth. It is deeply tied to Chinese national political and military power. "Shipyard infrastructure is clearly dual-use," they write. "This is why China's promotion and protection of advanced civilian facilities are likely related to its military goals." CSSC sees itself as fitting in this dual purpose, serving both civilian and military purposes. CSSC builds destroyers and other warships for the Chinese navy. As the company notes on its English website, "As for the naval ships, CSSC is capable of building almost various kinds of warship and auxiliary vessels as well as the related equipment for the Chinese Navy, thus earning the position as the backbone forces backing-up the Chinese navy in terms of its construction." CSSC builds destroyers for the Chinese navy at the Jiangnan shipyards as well as other surface combatants. The Chinese government has openly credited James Chao for his help in the expansion and growth of the Chinese shipbuilding industry. It is hard to underestimate the role that Chao has played in building the Chinese maritime business based on the laudatory statements made by senior Chinese government officials. As CSSC notes on its website, Chao has played an important role in advising Beijing on the development of its shipping business. No less than Chinese president Hu Jintao has thanked Chao "for his years of support to the Chinese shipbuilding industry and the contribution to the shipbuilding industry going into the international market." CSSC noted that Chao "ordered a good quantity of ships" from the state company and with his efforts "virtually pushed forward Chinese effort in opening the international market." Indeed, on November 9, 2006, James Chao went to the Great Hall of the People in Beijing where he was greeted by Chinese premier Wen Jiabao, Vice President Liu Xiaoyan, and Chen Xiaojin, the president of the CSSC. They thanked him for his contributions to Chinese shipbuilding. That is high praise given that the Chinese government sees the shipbuilding industry as a military asset. Beijing "has assigned the shipbuilding industry a key role in China's development as a

great power," writes Professor Bernard Cole of the National War College. Stated similarly elsewhere, the Chinese government has "assigned the [shipbuilding] industry a key role in supporting China's continued development as a great power." For Chinese maritime strategists, the ability to manufacture ships is largely about one threat: the U.S. Navy. "Beijing's 2015 military strategy," says Professor Cole, "clearly refers to the United States as China's most threatening security concern . . . Hence, the country must 'build a powerful navy.'"

56 Andrew S. Erickson, "Introduction: China's Military Shipbuilding Industry Steams Ahead, on What Course?" in Erickson, ed., *Chinese Naval Shipbuilding: An Ambitious and Uncertain Course*, 13.

57 Julian Snelder, "China's Civilian Shipbuilding in Competitive Context: An Asian Industrial Perspective," in Erickson, ed., *Chinese Naval Shipbuilding: An Ambitious and Uncertain Course*, 192.

58 Manu Raju and John Bresnahan, "Members' Fortunes See Steep Declines," *Politico*, June 12, 2009, http://www.politico.com/story/2009/06/members-fortunes-see-steep-declines-023693.

59 "Former U.S. Secretary of Labor Elaine Chao to Keynote HR Technology Conference & Expo in China," *Sys-Con Media*, February 9, 2016 http://news.sys-con.com/node/3671883; Sun Xiaojing, "Elaine L. Chao: The Wise with Grace," Peking University, accessed April 4, 2017, http://english.pku.edu.cn/News_Events/PointofView/922.htm; Associated Press, "Trump Appointee Spoke at Event for 'Cult-like' Iran Exile Group," *CBS News*, February 5, 2017, https://www.cbsnews.com/news/elaine-chao-trump-transportation-secretary-paid-iran-exile-group-mek-ap/.

60 "Boao Forum for Asia 2013 at a Glance," *ChinaDaily.com*, April 5, 2013; "Select List of Confirmed Speakers/Guests," Boao Forum for Asia's website, accessed April 4, 2017, http://english.boaoforum.org/ac2013slocsg/index.jhtml.

61 Angus Griggin, "You Know Who Your Friends Are at Boao," *Australian Financial Review*, April 6, 2013, Nexis; "Boao Forum for Asia 2013 at a Glance," *China Daily*, April 5, 2013; "Select List of Confirmed Speakers/Guests," Boao Forum for Asia website, accessed April 4, 2017; "Davos" is the location (Switzerland), and therefore a shorthand term for a yearly gathering sponsored by the World Economic Forum where business and political leaders around the globe come to discuss world economics. It also has a reputation as a party for the rich and famous.

62 Susan Brownell, "Another Side of the Shanghai World Expo: Forum on ICT and Urban Development," *China Beat*, June 9, 2010, http://www.thechinabeat.org/?p=2167.

63 Li Han, "U.S. Secretary of Labor Elaine L. Chao Speaks at Tsinghua," Tsinghua University News, accessed April 4, 2017, http://news.tsinghua.edu.cn/publish/thunewsen/9673/2011/20110111161219418957849/20110111161219418957849_.html; Liu Xiya, "James Si-Cheng Chao's Family Speak

at PKU," Peking University website, October 21, 2015, http://www.oir.pku.edu.cn/En/html/2015/NewsExpress_1021/331.html; Sun Xiaojing, "Elaine L. Chao: The Wise with Grace"; "E-Journal (October, 2009)," Fudan University website, November 23, 2009, http://www.fao.fudan.edu.cn/40/66/c1666a16486/page.htm#s7.

64 Judis, "Sullied Heritage."

65 "Trump Slams China for Currency Manipulation, IP Rights," *Today*, July 22, 2016, http://www.todayonline.com/world/americas/trump-slams-china-currency-manipulation-ip-rights.

66 Reena Flores, "Donald Trump Tweets About China's Currency, Military Policies After Taiwan Flap," *CBS News*, December 4, 2016, http://www.cbsnews.com/news/donald-trump-tweets-china-currency-military-policies-taiwan-flap/.

67 Michael F. Martin, *China's Banking System: Issues for Congress*, Congressional Research Service Report, February 20, 2012, 26.

68 "Directors and Board of Directors," Bank of China website, accessed April 5, 2017, http://www.bankofchina.com/en/investor/ir6/201504/t20150402_4830133.html?keywords=angela+chao.

CHAPTER 6: THE PRINCELINGS OF K STREET

1 "KLTZ/Mix-93 February 2000 News Archive," KLTZ News Archive, accessed May 15, 2017, http://www.kltz.com/archives/newsfeb00.html; "United States House of Representatives elections, 2000," Wikiwand, accessed May 15, 2017, https://www.wikiwand.com/en/United_States_House_of_Representatives_elections,_2000.

2 "Leadership," *gage.cc*, accessed April 22, 2017, http://gage.cc/about/executive-team/; David Morgan, "Tainted Ex-Senator Joins Lobbying Firm," *CBS News.com*, January 10, 2007, http://www.cbsnews.com/news/tainted-ex-senator-joins-lobbying-firm/.

3 LD-2 lobbying disclosure form for GAGE LLC for 2011, quarter 3, https://soprweb.senate.gov/index.cfm?event=getFilingDetails&filingID=01CBA292–132E-446D-ACD7–890C07F40A5B&filingTypeID=69.

4 LD-2 lobbying disclosure form for GAGE LLC for 2011, quarter 1, https://soprweb.senate.gov/index.cfm?event=getFilingDetails&filingID=44FF7300–7446–4A0E-BD47-E57D95195D7D&filingTypeID=51.

5 Matt Gouras, "Rehberg's Son Works for Lobbyists," Associated Press, October 18, 2011, http://www.deseretnews.com/article/700189286/APNewsBreak-Rehbergs-son-works-for-lobbyists.html?pg=all.

6 "Mongolia hires advocacy representation in Washington, DC," *gage.cc*, http://gage.cc/press-news/.

7 Matt Gouras, Associated Press, "Rehberg attacks lobbyists, while son represents Mongolian government in D.C.," *Missoulian*, October 19, 2011,

http://missoulian.com/news/local/rehberg-attacks-lobbyists-while-son-represents-mongolian-government-in-d/article_ba01d8bc-f9cd-11e0-b6d4–001cc4c002e0.html.

8 "Uranium War in Mongolia," May 5, 2010, *Mongolia-Web.com*, http://www.mongolia-web.com/2727-uranium-war-mongolia/.

9 LD-2 lobbying disclosure form for Gage LLC for 2011, quarter 1.

10 Gouras, "Rehberg's Son Works for Lobbyists"; "AJ Rehberg," profile, *Linkedin.com*, accessed October 1, 2016, https://www.linkedin.com/in/aj-rehberg-7069ab24.

11 Transparency International, "Corruption Perceptions Index 2013," *Transpoarency.org*, https://www.transparency.org/cpi2013/results.

12 "Mongolia Compact," Millenium Challenge Corporation, accessed May 16, 2017, https://www.mcc.gov/where-we-work/program/mongolia-compact.

13 State, Foreign Operations, And Related Programs Appropriations for 2010: Hearings before the Subcomm. On State, Foreign Operations, and Related Programs of the H. Comm. On Appropriations, 111th Cong. (2010) (statement of Nita M. Lowey, Chairwoman, New York) https://www.gpo.gov/fdsys/pkg/CHRG-111hhrg55951/html/CHRG-111hhrg55951.htm.

14 *Ballotpedia Online*, s.v. "Denny Rehberg," accessed May 16, 2017, https://ballotpedia.org/Denny_Rehberg.

15 Alissa Irei, "Rehberg Reportedly Spending Time with Family, in Days Following Senate Loss," *NBC Montana*, November 8, 2012, http://www.nbcmontana.com/news/politics/local-elections/rehberg-reportedly-spending-time-with-family-in-days-following-senate-loss/7868290; "Hon., Denny Rehberg," Mercury LLC website, accessed October 12, 2017, http://www.mercuryllc.com/experts/hon-denny-rehberg/.

16 Jasper Craven, "Leahy's Daughter Lobbies Senate for Hollywood," *VTDigger*, September 11, 2016, https://vtdigger.org/2016/09/11/leahys-daughter-lobbies-senate-hollywood/.

17 David S. Fallis and Dan Keating, "In Congress, Relatives Lobby for Bills Before Family Members," *Washington Post*, December 29, 2012, https://www.washingtonpost.com/investigations/in-congress-relatives-lobby-for-bills-before-family-members/2012/12/29/a54adce2–4301–11e2–9648-a2c323a991d6_story.html?utm_term=.c348bccb6741.

18 Jonathan Poet, "Bud Shuster to Retire," *ABC News*, January 4, 2001, http://abcnews.go.com/Politics/story?id=122122; Paul Kane, "Bill Shuster, Who Took Over His Father's District, Says 'It's a Different Congress' Today," *Washington Post*, May 19, 2014, https://www.washingtonpost.com/politics/bill-shuster-who-took-over-his-fathers-district-says-its-a-different-congress-today/2014/05/19/3de8b476-dc7e-11e3–8009–71de85b9c527_story.html?utm_term=.be438b93b287.

19 "Shuster to Chair House Transportation Committee," *Transport Topics*, November 28, 2012 http://www.ttnews.com/articles/basetemplate.aspx?storyid=30696&t=Shuster-to-Chair-House-Transportation-Committee.

20 Eugene Mulero and Michelle Merlin, "Not His Father's Committee—Shuster Tries to Boost Infrastructure Investments in a Time of Fiscal Austerity," *E&E*, April 10, 2013, https://www.eenews.net/stories/1059979150.

21 Ibid.

22 LD-2 lobbying disclosure form for Buchanan Ingersoll & Rooney PC for 2013, quarter 3, https://soprweb.senate.gov/index.cfm?event=getFilingDetails&filingID=35A833A9–0EF0–4854-B220-F4FF0BA17775&filingTypeID=69; LD-2 lobbying disclosure form for Buchanan Ingersoll & Rooney PC for 2013, quarter 4, https://soprweb.senate.gov/index.cfm?event=getFilingDetails&filingID=64243951–1D87–4A09-AD50-DA317C359F42&filingTypeID=78.

23 LD-2 lobbying disclosure form for Buchanan Ingersoll & Rooney PC for 2013, quarter 4, https://soprweb.senate.gov/index.cfm?event=getFilingDetails&filingID=922A246C-602D-4340–8D75–8918E2695B39&filingTypeID=78.

24 LD-2 lobbying disclosure form for Buchanan Ingersoll & Rooney PC for 2015, quarter 2, https://soprweb.senate.gov/index.cfm?event=getFilingDetails&filingID=09E6C495-A7CC-457A-ADD0–36109017AD36&filingTypeID=60.

25 Pipeline Safety, Regulatory Certainty, and Job Creation Act of 2011, H.R. 2845, 112th Cong. (2011), https://www.congress.gov/bill/112th-congress/house-bill/2845.

26 "Robert L. Shuster," on Buchanan Ingersoll & Rooney's website, accessed May 16, 2017, http://www.bipc.com/robert-shuster.

27 Open Secrets, "Firm Profile: Lobbyists 2017," Primerica Life Insurance Lobbyists, accessed October 12, 2017, https://www.opensecrets.org/lobby/firmslbs.php?id=D000020123&year=2017; "Allison Shuster," profile, LinkedIn, accessed May 16, 2017, https://www.linkedin.com/in/allison-shuster-712b0b2a/.

28 Tracie Mauriello, "Rep. Bill Shuster Admits to 'Personal Relationship' with Lobbyist," *Pittsburgh Post-Gazette*, April 17, 2015, http://www.post-gazette.com/news/nation/2015/04/17/Shuster/stories/201504170211.

29 Ibid.

30 An ethics panel in North Carolina dealt with this question and concluded, "Consensual sexual relationships do not have monetary value and therefore are not reportable as gifts or 'reportable expenditures made for lobbying.'" Olivia Nuzzi, "North Carolina Lobbyists Can Officially Screw Politicians Legally," *DailyBeast.com*, February 18, 2015, http://www.thedailybeast.com/north-carolina-lobbyists-can-officially-screw-politicians-legally.

31 Marin Cogan, "When Members of Congress Sleep With Lobbyists," *New York Magazine*, April 17, 2015, http://nymag.com/daily/intelligencer /2015/04/when-members-of-congress-sleep-with-lobbyists.html.

32 Marianne Bertrand, Matilde Bombardini, and Francesco Trebbi, "Is It Whom You Know or What You Know? An Empirical Assessment of the Lobbying Process," *American Economic Review* 104, no. 12 (December 2014): 3887.

33 Carl Hulse, "In Capitol, Last Names Link Some Leaders and Lobbyists," *New York Times*, August 4, 2002, http://www.nytimes.com/2002/08/04 /us/in-capitol-last-names-link-some-leaders-and-lobbyists.html; "Chester T. Lott," on Squire Patton Boggs's website, accessed May 16, 2017 http:// www.squirepattonboggs.com/en/professionals/l/lott-chester-trent.

34 Verified Complaint, Cadle v. Jefferson, No. 3:07CV-70-S (D.KY. 2007), http://hosted.ap.org/specials/interactives/_documents/jefferson_suit.pdf; Neil A. Lewis, "Jury Prepares for Jefferson Bribery Case," *New York Times*, July 29, 2009 http://www.nytimes.com/2009/07/30/us/politics/30 jefferson.html?_r=0.

35 Ibid.

36 Ibid.; Lewis, "Jury Prepares for Jefferson Bribery Case"; Allan Lengel, "Nigerian Entangled in Jefferson Investigation," *Washington Post*, July 22, 2006, http://www.washingtonpost.com/wp-dyn/content/article/2006/07/21 /AR2006072101536_pf.html.

37 Verified Complaint, Cadle v. Jefferson.

38 U.S. Department of Justice, "Former Congressman William J. Jefferson Sentenced to 13 Years in Prison for Bribery and Other Charges," press release, November 13, 2009, https://www.justice.gov/opa/pr/former -congressman-william-j-jefferson-sentenced-13-years-prison-bribery -and-other-charges.

39 "The Making of the 'Loretta Sanchez Scandal,'" *Latino Politics* (blog), accessed May 16, 2017 http://latinopoliticsblog.com/2009/06/02/the -making-of-the-"loretta-sanchez-scandal"/; "Love, etc.: Rep. Loretta Sanchez Weds Jack Einwechter; Mamie Gummer Marries Ben Walker," *Washington Post*, July 17, 2011, https://www.washingtonpost.com/blogs /reliable-source/post/love-etc-rep-loretta-sanchez-weds-jack-einwechter -mamie-gummer-marries-ben-walker/2011/07/17/glQAhW1aKI_blog .html?utm_term=.c9e3cc2ff783.

40 Open Secrets, "Einwechter, John P.," Lobbyist Profile: Summary, 2007, accessed May 16, 2017 https://www.opensecrets.org/lobby/lobbyist .php?id=Y0000000259L&year=2007; Open Secrets, "Einwechter, John P.," Lobbyist Profile: Summary, 2008, accessed May 16, 2017; https://www .opensecrets.org/lobby/lobbyist.php?id=Y0000000259L&year=2008; Open Secrets, "Einwechter, John P.," Lobbyist Profile: Summary, 2009, accessed May 16, 2017, https://www.opensecrets.org/lobby/lobbyist.php?id=Y0000 000259L&year=2009.

41 "Contacts for Greenberg Traurig LLP; Kurdistan Regional Government (Previously Known as Kurdish Democratic Party)," Foreign Influence Explorer, accessed May 16, 2017, http://foreign.influenceexplorer.com/contact-table?reg_id=5712&client_id=129953&p=7; Marisa M. Kashino, "Flying the Flag," *Washingtonian*, April 1, 2010, https://www.washingtonian.com/2010/04/01/flying-the-flag/; Report of the Attorney General to the Congress of the United States on the Administration of the Foreign Agents Registration Act of 1938, as amended, for the six months ending June 30, 2011, U.S. Department of Justice, n.d., accessed October 24, 2017, https://www.fara.gov/reports/FARA_SAR_063011.pdf.

42 "The Making of the 'Loretta Sanchez Scandal.'"

43 "Curt Weldon," AFO Research, accessed May 16, 2017, http://www.aforeearch.com/curt-weldon.html; Ken Silverstein, Chuck Neubauer, and Richard T. Cooper, "Lucrative Deals for a Daughter of Politics," *Los Angeles Times*, February 20, 2004, http://articles.latimes.com/2004/feb/20/nation/na-weldon20.

44 Ibid.; "FBI raid in Weldon probe spurs more questions than answers," The Mercury columns, October 19, 2006, http://www.pottsmerc.com/article/MP/20061019/OPINION03/310199986; Richard B. Schmitt and Chuck Neubauer, "FBI Raids Lobbyists' Homes, Office," *Los Angeles Times*, October 17, 2006, http://articles.latimes.com/2006/oct/17/nation/na-weldon17; "Osprey Officer Backs Out," *CBSNews.com*, June 27, 2000, http://www.cbsnews.com/news/osprey-officer-backs-out/.

45 Silverstein, Neubauer, and Cooper, "Lucrative Deals for a Daughter of Politics."

46 Ibid.

47 Glenn R. Simpson, "U.S. Probes Possible Crime Links to Russian Natural-Gas Deals," *Wall Street Journal*, December 22, 2006, https://www.wsj.com/articles/SB116675522912457466; "Russian Rags-to-riches Story," *Jacksonville.com*, June 25, 1998, http://jacksonville.com/tu-online/stories/062598/met_2A1itera.html#.WdaUzDBryUI.

48 Silverstein, Neubauer, and Cooper, "Lucrative Deals for a Daughter of Politics."

49 Ibid.

50 Ibid.

51 Phil Heron, "The Heron's Nest: A Full Spectrum of Memories," *Daily Times News*, July 16, 2008, http://www.delcotimes.com/article/DC/20080716/NEWS/307169991; Carol D. Leonnig and R. Jeffrey Smith, "Homes Raided in Rep. Weldon Influence Probe FBI Looks at Business Run by Daughter, Political Ally," *Washington Post*, October 17, 2006, https://www.washingtonpost.com/archive/politics/2006/10/17/homes-raided-in-rep-weldon-influence-probe-span-classbankheadfbi-looks-at-business-run-by-daughter-political-allyspan/bf6ee318-8742-45ff-8d5a-54d4a5834943/?utm_term=.4a97a4e6d0cb.

52 Kirk Johnson, "Leader of '76 Insurgency Is Now a Target of One," *New York Times*, May 22, 2012, http://www.nytimes.com/2012/05/23/us/politics/hatch-finds-76-tactics-now-used-against-him.html; "Biography," on Senator Orrin Hatch's website, accessed May 16, 2017, https://www.hatch.senate.gov/public/index.cfm/biography.

53 Chuck Neubauer, Judy Pasternak, and Richard T. Cooper, "A Washington Bouquet: Hire a Lawmaker's Kid," *Los Angeles Times*, June 22, 2003, http://articles.latimes.com/2003/jun/22/nation/na-sons22.

54 "Can Shering-Plough's Aggressive Lobbying Buy a Claritin Patent Extension?" *Public Citizen*, July 30, 1999, https://www.citizen.org/article/can-schering-ploughs-aggressive-lobbying-buy-claritin-patent-extension.

55 Holly Bailey, "Gore's Legal Fees, Tax Cuts and Asbestos Lobbying: Money in Politics News Briefs," *Center for Responsive Politics*, 6 no. 8 (February 2001), on Internet archive, accessed June 13, 2016, http://web.archive.org/web/20011110104234/;

56 Chuck Neubauer, Judy Pasternak, and Richard T. Cooper, "Senator, His Son Get Boosts for Makers of Ephedra," *Los Angeles Times*, March 5, 2003.

57 Ibid.

58 *Source Watch*, "Walker Martin & Hatch," https://www.sourcewatch.org/index.php/Walker_Martin_%26_Hatch.

59 Neubauer, Pasternak, and Cooper, "Senator, His Son Get Boosts for Makers of Ephedra."

60 Neubauer, Pasternak, and Cooper, "A Washington Bouquet: Hire a Lawmaker's Kid"; Scott D. Pierce, "HBO Report Links Orrin Hatch's Dietary Supplement Legislation to Military Fatalities," *Salt Lake Tribune*, May 19, 2015, http://www.sltrib.com/blogs/tv/2530492–155/story.html.

61 Suzanne Struglinski, "Hatch Votes 'Present' to Avoid Conflict," *Deseret News*, December 21, 2007, http://www.deseretnews.com/article/695237929/Hatch-votes-present-to-avoid-conflict.html?pg=all.

62 Ibid.

63 Cogan, "When Members of Congress Sleep With Lobbyists."

64 "Durbin, Richard Joseph," Biographical Directory of the United States Congress, accessed October 12, 2017, https://www.durbin.senate.gov/about-dick-durbin.

65 Katherine Skiba and Kim Geiger, "When Interests Overlap for Durbin, Lobbyist Wife," *Chicago Tribune*, October 4, 2014, http://www.chicagotribune.com/news/ct-durbin-wife-lobbyist-20141004-story.html.

66 Ibid.

67 Selma Lussenburg, Michael Barrett, and Paul Durbin, "Finding the Money: Securing Capital for Energy Innovation," *Canada-United States Law Journal* 36, no. 2 (January 2012): 282, Ebsco.

68 Energy and Water Development Subcomm. Webpage, U.S. Comm. On Appropriations, access May 17, 2017, https://www.appropriations.senate.gov/subcommittees/energy-and-water-development.

69 "About Marty Durbin," on Energy API's website, accessed May 17, 2017, http://www.api.org/about/marty-durbin.

70 Steve Horn, "The Fracking Lobby Groups and America's Natural Gas Alliance (ANGA), Public Relations Arm of the Fracking Industry," Global Research, November 20, 2013, http://www.globalresearch.ca/the-fracking-lobby-groups-and-americas-natural-gas-alliance-anga-public-relations-arm-of-the-fracking-industry/5358870.

71 Lee Fang, "The Fracking Candidate: It's All in the Family for Rep. Shelley Moore Capito," *National Memo*, June 25, 2014, http://www.nationalmemo.com/fracking-candidate-family-rep-shelley-moore-capito/.

72 Anna Palmer and Darren Goode, "Sources: Oil, Natural Gas Lobby Groups in Merger Talks," *Politico*, September 4, 2015, http://www.politico.com/story/2015/09/american-petroleum-institute-americas-natural-gas-alliance-merger-talks-213356.

73 Craven, "Leahy's Daughter Lobbies Senate," https://vtdigger.org/2016/09/11/leahys-daughter-lobbies-senate-hollywood/.

CHAPTER 7: THE PRINCELINGS OF CHICAGO

1 Evan Osnos, "The Daley Show," *New Yorker*, March 8, 2010, http://www.newyorker.com/magazine/2010/03/08/the-daley-show; Patrick Reardon, "A.J. Liebling's classic 'Chicago: The Second City' put Chicago in its place," *Chicago Tribune*, May 30, 2016, http://www.chicagotribune.com/lifestyles/books/ct-prj-chicago-the-second-city-20160503-story.html.

2 "Remembering Richard J. Daley," Special Collections and University Archives, 2015, http://rjd.library.uic.edu/biographies/richard-j-daley/.

3 "House of Lords," *Investor's Business Daily*, September 7, 2004, ProQuest.

4 Osnos, "The Daley Show."

5 Andrew Stevens, "Richard M Daley Former Mayor of Chicago," City Mayors website, May 2011, http://www.citymayors.com/mayors/chicago_mayor.html.

6 Osnos, "The Daley Show."

7 Costas Spirou and Dennis R. Judd, *Building the City of Spectacle: Mayor Richard M. Daley and the Remaking of Chicago* (Ithaca, NY: Cornell University Press, 2016), 154.

8 Osnos, "The Daley Show"; Spirou and Judd, *Building the City of Spectacle*, 155.

9 Ibid., 156.

10 Keith Koeneman, *First Son: The Biography of Richard M. Daley* (Chicago: University of Chicago Press, 2013), 146; Peter Kendall, "Teen Injured at Daley's Home Improves," *Chicago Tribune*, March 4, 1992, http://articles.chicagotribune.com/1992–03–04/news/9201200738_1_mayor-richard-j-daley-beer-cans-bat; Peter Kendall, "Daley's Son Pleads Guilty in Brawl,"

Chicago Tribune, April 11, 1992, http://articles.chicagotribune.com/1992–04–11/news/9202020408_1_daley-family-wiley-mayor-richard-daley.

11 "Richard M. Daley's 22 Years as Mayor," *Chicago Tribune*, April 30, 2011, http://www.chicagotribune.com/news/ct-met-daley-timeline-special-section20110430-story.html.

12 Kendall, "Daley's Son Pleads Guilty in Brawl."

13 George Papajohn, "Final Charges Are Filed in Fight at Daley Retreat," *Chicago Tribune*, April 25, 1992, http://articles.chicagotribune.com/1992–04–25/news/9202060753_1_youths-mayor-richard-daley-daley-home.

14 Spirou and Judd, *Building the City of Spectacle*, 155.

15 Hal Dardick and Jeff Coen, "Former Business Partner of Daley's Son Indicted," *Chicago Tribune*, January 6, 2011, http://articles.chicagotribune.com/2011–01–06/news/ct-met-city-minority-fraud-20110106_1_municipal-sewer-services-patrick-daley-robert-vanecko.

16 Gary Washburn, "Mayor Calls Son's Deal a 'Lapse in Judgment,'" *Chicago Tribune*, December 19, 2007, http://articles.chicagotribune.com/2007–12–19/news/0712180755_1_mayor-richard-daley-navy-pier-contract.

17 Dardick and Coen, "Former Business Partner of Daley's Son Indicted"; "Rauner-Daley Connection," *Chicago Sun-Times*, July 14, 2014, https://www.pressreader.com/usa/chicago-sun-times/20140714/281638188317939.

18 Tim Novak and Chris Fusco, "Former Mayor Daley's Son Profited After Airport Wi-Fi Deal," *Chicago Sun-Times*, June 6, 2011, http://archives.californiaaviation.org/airport/msg47582.html.

19 "Feds Take Control of VC Fund That Paid Daley Son $1.2 Mil.: Report," *Crain's Chicago Business*, June 27, 2011, http://www.chicagobusiness.com/article/20110627/NEWS02/110629883/feds-take-control-of-vc-fund-that-paid-daley-son-1–2-mil-report.

20 "Russian Émigré Now Chicago's CAB King; A Friend of Daley's Son," *Chicago Sun-Times*, September 29, 2014, http://chicago.suntimes.com/politics/russian-emigre-now-chicagos-cab-king-a-friend-of-daleys-son/.

21 Kathy Bergen, "Hard Driver at the Top of City's Heated Taxi Market," *Chicago Tribune*, May 10, 2004, http://articles.chicagotribune.com/2004–05–10/news/0405100182_1_world-business-chicago-young-driver-cab.

22 "Russian Stock Fraud Investigator Killed in Moscow," U.S. Department of the Treasury, Moscow, Russia, *WikiLeaks*, September 28, 2007, https://wikileaks.org/plusd/cables/07MOSCOW4768_a.html.

23 Ibid.

24 "Russian Émigré Now Chicago's CAB King; A Friend of Daley's Son."

25 Ibid.

26 Ibid.

27 Spirou and Judd, *Building the City of Spectacle*, 156.

28 Ned Levin, Emily Glazer, and Christopher M. Matthews, "In J.P. Morgan Emails, a Tale of China and Connections," *Wall Street Journal*, February 6,

2015, https://www.wsj.com/articles/in-j-p-morgan-emails-a-tale-of-china-and-connections-1423241289.

29 Ibid.

30 Ibid.

31 "William Daley, New Chief of Staff, Fainted Last Time He Got a White House Appointment," *Huffington Post*, January 7, 2011, http://www.huffingtonpost.com/2011/01/07/william-daley-new-chief-o_n_805870.html.

32 Carol Felsenthal, "The Truth About Valerie Jarrett, Mystery Woman of the White House," *Chicago Magazine*, January 31, 2014, http://www.chicagomag.com/Chicago-Magazine/Felsenthal-Files/January-2014/The-Mysteries-and-Realities-of-Valerie-Jarrett-Mystery-Woman-of-the-White-House/.

33 "The White House Daley Show," *Investor's Business Daily*, January 7, 2011, https://www.investors.com/politics/editorials/the-white-house-daley-show/.

34 Koeneman, *First Son*, 275.

35 Steven Sloan, "Bankers Tapped for Transition," *American Banker* 173, no. 215 (2008): 2, EBSCOhost.

36 Koeneman, *First Son*, 277.

37 Kathy Bergen and Jessica Wohl, "Russia Turmoil Begins to Bruise Illinois Businesses," *Chicago Tribune*, November 14, 2014, http://www.chicagotribune.com/business/ct-russia-sanctions-illinois-business-1116-biz-20141114-story.html.

38 "Tur Partners Asia Limited," HKG Business website, March 4, 2013, https://www.hkgbusiness.com/en/company/Tur-Partners-Asia-Limited; "Search Cyprus Companies," i-Cyprus website, accessed March 13, 2017, https://i-cyprus.com/search?q=Tur+Partners.

39 Liz Alderman, "Russians Return to Cyprus, a Favorite Tax Haven," *New York Times*, February 17, 2014, https://www.nytimes.com/2014/02/18/business/international/russian-business-target-of-cypriot-bailout-still-loves-the-island.html?_r=1.

40 "Team Page," Tur Partners website, accessed March 13, 2017, http://www.turpartners.com/team.html.

41 "Tur Partners Eurasia Ltd," i-Cyprus website, accessed March 13, 2017, https://i-cyprus.com/company/466008.

42 "Russian President Vladimir Putin Has Awarded the Order of Friendship to Mukharbek Aushev Deputy Head of the State Duma," Alamy, photograph, accessed March 13, 2017, http://www.alamy.com/stock-photo-russian-president-vladimir-putin-left-has-awarded-the-order-of-friendship-22976789.html.

43 Russian Direct Investment Fund, "Russian Direct Investment Fund Announces International Advisory Board," news release, September 16, 2011, https://rdif.ru/Eng_fullNews/53/.

44 Russian Direct Investment Fund, "RDIF Leads Consortium to Invest in Russia's Largest Agriculture Tire Producer," press release, June 19, 2013, https://rdif.ru/Eng_fullNews/251/.

45 Deanna Bellandi and Caryn Rousseau, "Hu Visit Shows China-Chicago Ties," January 22, 2011, Questia, http://www.questia.com/read/1G1–24723 4781/hu-visit-shows-china-chicago-ties.

46 "Team Page," Tur Partners website, accessed March 13, 2017, http://turpartners.com/team.html.

47 Erin Ailworth, "A123 Buyer Quietly Built Presence," *Boston Globe*, December 13, 2012, https://www.bostonglobe.com/business/2012/12/13/new-chinese-owner-has-american-roots/E5Pxs7EsocCsnJiVwyqFjN/story.html.

48 Office of Mgmt. & Budget, OMB No. 1124–000, Reg. No. 6196, Exhibit A to Registration Statement Pursuant to the Foreign Agents Registration Act of 1938, as amended, for Tur Partners LLC (November 18, 2013), https://www.fara.gov/docs/6196-Exhibit-AB-20131118–1.pdf.

49 Chen Weihua, "Liu Meets Kerry, Biden in DC," *ChinaDaily.com*, November 22, 2013, http://usa.chinadaily.com.cn/epaper/2013–11/22/content_17124444.htm.

50 For a general overview of the logistics structure, *Wikipedia*, s.v. "NATO Logistics in the Afghan War," last modified September 5, 2017, https://en.m.wikipedia.org/wiki/NATO_logistics_in_the_Afghan_War#Northern_Distribution_Network.

51 "Communists Won't Let NATO in Lenin's Birthplace," *RT*, April 4, 2012, https://www.rt.com/politics/ulyanovsk-nato-communists-protest-203/.

52 Dmitri Volkogonov, *Lenin*, (New York: Free Press, 1994), https://books.google.com/books?id=4Sz4VOBMm2QC&pg=PT30&lpg=PT30&dq=simbirsk+ulyanovsk+lenin&source=bl&ots=Nlv2uNp4Y-&sig=6dYYjeooSvSrHymbZJmockp9eVM&hl=en&sa=X&ved=0ahUKEwjB1Pvz0JLTAhUFYyYKHU4wB5w4ChDoAQgZMAA#v=onepage&q=simbirsk%20ulyanovsk%20lenin&f=false.

53 "ARR Corporation Will Create an Aircraft Maintenance Center in Ulyanovsk," *Russian Aviation*, March 5, 2012, http://www.ruaviation.com/news/2012/3/5/830/print/?h.

54 "President Vladimir Putin Instructed to Engage in the Company 'AAR Rus' in Ulyanovsk," *ULBusiness*, accessed on Google Translate, September 22, 2015, https://translate.google.com/translate?hl=en&sl=ru&u=http://ulbusiness.ru/vladimir-putin-poruchil-aar-rus-v-ulyanovske/&prev=search.

55 Matthew Mosk and Brian Ross, "The $500,000 Green Card: Obama, Clinton Kin Courted by Foreign Middlemen," *ABC News*, February 4, 2015, http://abcnews.go.com/International/500000-green-card-clinton-obama-kin-courted-foreign/story?id=28723441.

56 Tim Novak, "Watchdogs: Former Mayor Daley Hoping to Cash In on Green Cards," *Chicago Sun-Times*, June 24, 2016, http://chicago.suntimes.com/news/watchdogs-former-mayor-daley-hoping-to-cash-in-on-green-cards/.

CHAPTER 8: THE HYESAN YOUTH COPPER MINE OF NORTH KOREA

1 Russell Flannery, "Lu's U-Joints, and More," *Forbes*, March 28, 2008, https://www.forbes.com/global/2008/0407/070.html; William C. Kirby, "China's Prosperous Age: A Century in the Making," *China Heritage Quarterly*, no. 26 (June 2011), http://chinaheritagequarterly.org/features.php?searchterm=026_kirby.inc&issue=026.

2 Lynn T. White, *Political Booms: Local Money and Power in Taiwan, East China, Thailand, and the Philippines* (Hackensack, NJ: World Scientific Publishing, 2009), 148.

3 Ailworth, "A123 Buyer Quietly Built Presence."

4 Terence Corcoran, "Obama's Fisker Cliff; Battle of U.S. S.O.E. vs. Chinese S.O.E.," *National Post*, December 11, 2012, ProQuest.

5 Kirby, "China's Prosperous Age."

6 "Lu Guanqiu & Family," profile, *Forbes*, last modified June 13, 2017, https://www.forbes.com/profile/lu-guanqiu/.

7 Corcoran, "Obama's Fisker Cliff."

8 "Bush's Brother Becomes Counselor of Chinese Township Enterprise," Xinhua News Agency, October 28, 1999, Nexis.

9 "Bush vs. Forbes on China," *Weekly Standard*, November 22, 1999, http://www.weeklystandard.com/bush-vs.-forbes-on-china/article/11741.

10 Qilai Shen, "Chicago Mayor Richard Daley Visits Wanxiang Auto Parts (4 of 10), *Getty Images*, photograph, accessed May 17, 2017, http://www.gettyimages.com/event/chicago-mayor-richard-daley-visits-wanxiang-auto-parts-110827145?#lu-guanqiu-chairman-of-wanxiang-group-center-greets-richard-daley-of-picture-id110859751; Qilai Shen, "Chicago Mayor Richard Daley Visits Wanxiang Auto Parts (2 of 10)," *Getty Images*, photograph, accessed May 17, 2017, http://www.gettyimages.com/event/chicago-mayor-richard-daley-visits-wanxiang-auto-parts-110827145?#red-carpet-is-rolled-out-at-wanxiang-groups-electric-automobile-in-picture-id110859789.

11 Ailworth, "A123 buyer quietly built presence."

12 Rick Pearson and Bob Secter, "Obama and Daley Political Allies Now, But They're Hardly Cronies," *Chicago Tribune*, November 16, 2008, http://www.chicagotribune.com/news/chi-obama-daley17nov17-story.html.

13 Liu Chang, "Wanxiang Finds Success in US," *ChinaDaily.com*, July 28, 2014, http://usa.chinadaily.com.cn/2014–07/28/content_17944873.htm.

14 Nadja Brandt, "Wanxiang Team with Pritzker for $1 Billion in Hotels," *Bloomberg*, August 14, 2015, https://www.bloomberg.com/news/articles/2015–08–14/wanxiang-teams-with-pritzker-for-1-billion-in-hotels.

15 Kelvin Chan, "Survey: Indian Firms Best, China Worst on Transparency," Associated Press, July 11, 2016.

16 Dexter Roberts, "China's Companies Most at Risk for Corruption, Says Transparency International," *Bloomberg*, October 17, 2013, https://www.bloomberg.com/news/articles/2013-10-17/corruption-chinese-companies-most-at-risk-transparency-international-says.

17 Ana Ablaza, "Survey: Chinese Companies Most Corrupt Globally," *Yibada*, July 15, 2016, en.yibada.com.

18 Martin LaMonica, "Obama Signs Stimulus Plan, Touts Clean Energy," *CNET*, February 25, 2009, https://www.cnet.com/news/obama-signs-stimulus-plan-touts-clean-energy/.

19 "About Us," A123 Systems, accessed May 17, 2017, http://www.a123systems.com/about-us.htm.

20 "A123 Systems Awarded $249M Grant from U.S. Department of Energy to Build Advanced Battery Production Facilities in the United States," A123 Systems, press release, August 5, 2009, http://www.a123systems.com/567491e6-ed98-429f-b4a6-3ae3df789b90/media-room-2009-press-releases-detail.htm.

21 Tom Gara, "Obama's A123 Call, Coming to a Debate Near You," *Corporate Intelligence* (blog) *Wall Street Journal*, October 16, 2012, https://blogs.wsj.com/corporate-intelligence/2012/10/16/obamas-a123-call-coming-to-a-debate-near-you/.

22 Bill Vlasic and Matthew L. Wald, "Maker of Batteries Files for Bankruptcy," *New York Times*, October 16, 2012, http://www.nytimes.com/2012/10/17/business/battery-maker-a123-systems-files-for-bankruptcy.html.

23 Patrick Fitzgerald, "Chinese Firm Wins Auction for U.S.-Backed Battery Maker," *Wall Street Journal*, December 9, 2012, https://www.wsj.com/articles/SB10001424127887324024004578169653323996378.

24 Mark Pfeifle, "China's Strategic Next Step: Buying Critical US Taxpayer Funded Technology," The Blog (blog), *Huffington Post*, January 17, 2013, http://www.huffingtonpost.com/mark-pfeifle/china-military-technology_b_2473457.html.

25 Ayesha Rascoe, "Chinese Bid for A123 May Raise Security Risks: Senators," Reuters, November 30, 2012, http://www.reuters.com/article/us-usa-a123-security-idUSBRE8B000920121201.

26 Ibid.; "Understanding the CFIUS Process," Organization for International Investment, accessed May 17, 2017, http://www.ofii.org/sites/default/files/OFII_CFIUS_Primer.pdf.

27 "Lu Guanqiu," *Forbes*, photograph, January 19, 2011, https://www.forbes.com/pictures/mil45jgig/lu-guanqiu-2/#72cb1f6a49b7.

28 "Wanxiang America Opens Solar Panel Manufacturing Plant in Illinois," *Reliable Plant*, accessed May 17, 2017, http://www.reliableplant.com/Read/26188/Wanxiang-America-panel-plant.

29 Ailworth, "A123 Buyer Built Presence."

30 U.S. Federal Election Commission, "DNC Services Corp./Dem. Nat'l Committee," Schedule A (FEC Form 3x) Itemized Receipts (page 1648 of 2902), May 12, 2015, 11 C.F.R. 104.3 (2007), http://docquery.fec.gov/cgi-bin/fecimg/?15951508532.

31 Campaign Money, "Weidi Lu," political campaign contributions profile, 2016, accessed June 14, 2017, https://www.campaignmoney.com/political/contributions/weidi-lu.asp?cycle=16.

32 Corcoran, "Obama's Fisker Cliff."

33 Bill Vlasic, "Chinese Firm Wins Bid for Auto Battery Maker," *New York Times*, December 9, 2012, http://www.nytimes.com/2012/12/10/business/global/auction-for-a123-systems-won-by-wanxiang-group-of-china.html.

34 Mark Pfeifle, "China's Strategic Next Step."

35 Policy and Regulatory Report, "A123/Wanxiang: CFIUS Mitigation Labeled a 'Technical Fiction' by Advocacy Group," January 14, 2012, https://www.sheppardmullin.com/media/article/1269_A123%20Wanxiang%20PaRR%2015%20January%202013.pdf.

36 Ibid.

37 Theodore H. Moran, "Chinese Investment and CFIUS: Time for an Updated (and Revised) Perspective," policy brief, *Peterson Institute for International Economics*, no. 15–17, Washington, DC (September 2015), https://piie.com/publications/pb/pb15–17.pdf.

38 "Council Responds to Approval of A123 Systems Sale to Shanghai-Based Wanxiang Group," Strategic Minerals Advisory Council, January 29, 2013, http://www.strategicmaterials.org/2013/01/29/council-responds-to-approval-of-a123-systems-sale-to-shanghai-based-wanxiang-group/.

39 Ibid.

40 "Battery Maker A123 of China's Wanxiang Ressurects from Bankruptcy," *ChinaDaily.com*, July 9, 2014, http://www.chinadaily.com.cn/business/motoring/2014–07/09/content_17688946.htm.

41 David Veselenak, "A123 Systems Relocating to Novi, Slashing 200 Jobs," *Detroit Free Press*, April 4, 2017, http://www.freep.com/story/money/cars/2017/04/04/a123-systems-moving-novi/100020820/.

42 "China-DPRK Copper Joint Venture Starts Operation," *Xinhua News Agency*, September 19, 2011, http://news.xinhuanet.com/english2010/china/2011–09/19/c_131147696.htm.

43 "Mainland Companies Buy Stake in North Korea's biggest Copper Mine," *Nation*, December 26, 2006, https://www.pressreader.com/thailand/the-nation/20061226/282273840880756.

44 J.R. Mailey, *Hiding in Plain Sight: Cowboys, Conmen and North Korea's $6 Trillion Natural Resource Prize*, 38 North and US-Korea Institute at SAIS (2016), http://38north.org/wp-content/uploads/2016/04/201604_HIDING-IN-PLAIN-SIGHT_JRMailey.pdf; U.S. Department of the Treasury, "Treasury Imposes Sanctions Against the Government of the Democratic People's

Republic of Korea," news release, January 2, 2015, https://www.treasury.gov/press-center/press-releases/Pages/jl9733.aspx; Mailey, *Hiding in Plain Sight*.

45 Lin Shi, *The Mineral Industry of North Korea*, prepared by the U.S. Geological Survey Minerals Yearbook in cooperation with the U.S. Department of the Interior (March 2014), https://minerals.usgs.gov/minerals/pubs/country/2012/myb3–2012-kn.pdf.

46 Dexter Roberts, "North Korea, New Land of Opportunity?" *Bloomberg*, January 19, 2012, https://www.bloomberg.com/news/articles/2012–01–19/north-korea-new-land-of-opportunity.

47 Carla P. Freeman, *China and North Korea: Strategic and Policy Perspectives from a Changing China* (London: Palgrave Macmillan, 2015), https://books.google.com/books?id=XZBMCgAAQBAJ&pg=PT224&lpg=PT224&dq=Hyesan+Youth+Mine&source=bl&ots=F07t1FXRox&sig=PakQaoDwHW4qcBTMNMwtBYjMco4&hl=en&sa=X&ved=0ahUKEwiJ4Kaio8XTAhWDLSYKHe6BCqMQ6AEIZjAN#v=onepage&q=Hyesan%20Youth%20Mine&f=false.

48 Moon Sung Hwi, "Ex-Soldiers Bolt Copper Mine," trans. Grigore Scarlatoiu, ed. Richard Finney, *Radio Free Asia*, November 18, 2010, http://www.rfa.org/english/news/korea/assignments-11182010145704.html.

49 "Milestones," on Wanxiang Resources' website, accessed June 13, 2017, http://www.wxresources.com/en/about/GreatEvent.aspx; "China-DPRK Copper Joint Venture Starts Operation"; "China-N. Korea JV Starts Production at Copper Mine," Reuters, September 20,2011, http://af.reuters.com/article/commoditiesNews/idAFL3E7KK0CZ20110920; "Chinese Joint Venture Company Takes Over Hyesan Youth Copper Mine," *North Korean Economy Watch*, accessed June 13, 2017, http://www.nkeconwatch.com/2011/09/19/chinese-joint-venture-company-takes-over-hyesan-youth-copper-mine/. Note: this last link is the source for the previous two. The link to the KCNA site is inoperable, so the information is only contained in the excerpt on this site, http://www.koreaherald.com/view.php?ud=20161130000971.

50 *China's Impact on Korean Peninsula Unification and Questions for the Senate*, A Minority Staff Report, Committee on Foreign Relations, United States Senate, December 11, 2012, 9, https://www.gpo.gov/fdsys/pkg/CPRT-112SPRT77566/pdf/CPRT-112SPRT77566.pdf. The report notes: "In 2007, the DPRK Ministry of Mining Industries and the Wanxiang Resources Limited Company of China established Hyesan-China Joint Venture Mineral Company. Copper ore production from the Hyesan Youth Mine goes to China. North Korean and Chinese workers completed a modernization project of the mine in September of 2011. The Hyesan Youth Mine reportedly has an annual capacity of 50 to 70,000 tons of copper concentrate that is expected to contain 20–30 percent copper with all of it to be sold

to China. Also, China, has reportedly invested $860 million in the mines under this joint venture and holds a 51 percent stake." Ibid., 15.

51 Shi, *The Mineral Industry.*

52 Stephen Haggard, "Ripping Off Foreigners II: China Edition," Peterson Institute for International Economics, October 24, 2012, https://piie.com/blogs/north-korea-witness-transformation/ripping-foreigners-ii-china-edition; "Northeast Asia Cooperative Security Project, North Korea Chronology 2012," *Social Science Research Council* (January 2012): 2, http://webarchive.ssrc.org/NK/NKCHRON%202012.pdf, 2.

53 Roberts, "North Korea, New Land of Opportunity?"

54 Richard Roth, Holly Yan, and Ralph Ellis, "U.N. Security Council Approves Tough Sanctions on North Korea," *CNN.com*, March 3, 2016, http://www.cnn.com/2016/03/02/world/un-north-korea-sanctions-vote/; "UN Tightens Sanctions on North Korea After Largest Nuclear Test Yet," *Guardian*, November 30, 2016, https://www.theguardian.com/world/2016/nov/30/un-north-korea-sanctions-nuclear-test.

55 Exec. Order No. 13,722,81 C.F.R. 14941 (2016), https://obamawhitehouse.archives.gov/the-press-office/2016/03/16/executive-order-blocking-property-government-north-korea-and-workers.

56 "Sanctions Aimed at Cutting North Korean Coal Exports," *Korea Herald*, November 30, 2016, http://www.koreaherald.com/view.php?ud=20161130000971.

57 U.S. Department of Justice, "Four Chinese Nationals and China-Based Company Charged with Using Front Companies to Evade U.S. Sanctions Targeting North Korea's Nuclear Weapons and Ballistic Missile Programs," news release, September 26, 2016, https://www.justice.gov/opa/pr/four-chinese-nationals-and-china-based-company-charged-using-front-companies-evade-us.

58 "Wanxiang Is Largest Importer of N.K. Minerals: Radio Free Asia," *Dong-A Ilbo*, October 6, 2016, http://english.donga.com/Home/3/all/26/755518/1.

59 Elizabeth Shim, "Chinese Company Has Exclusive Rights to North Korea Mineral Deposits," *United Press International*, October 5, 2016, http://www.upi.com/Top_News/World-News/2016/10/05/Report-Chinese-company-has-exclusive-rights-to-North-Korea-mineral-deposits/9801475686795/?spt=sec&or=tn.

CHAPTER 9: BARACK OBAMA'S BEST FRIEND

1 Deanna Bellandi, "Capital Culture: Even Barack Obama Needs a BFF," Associated Press, February 23, 2010, http://archive.boston.com/news/nation/articles/2010/02/23/capital_culture_even_barack_obama_needs_a_bff/.

2 "Obama Kicks Back with 'FOB#1' (Friend of Barack): First Family Enjoys Hawaiian Vacation with President's Best Buddy," *Daily Mail*, December 28,

2011, http://www.dailymail.co.uk/news/article-2079256/Obama-girls-visit-popular-snorkeling-spot-marine-sanctuary.html.

3 Melissa Harris, "Executive Profile: Martin Nesbitt, the First Friend," *Chicago Tribune*, January 21, 2013, http://www.chicagotribune.com/business/ct-biz-0121-executive-profile-nesbitt-20130121-story.html.

4 Ibid.

5 Ibid.

6 Michael Stratford and Kimberly Hefling, "Bid to Buy For-Profit College by Former Obama Insiders Raises Questions," *Politico*, June 29, 2016, http://www.politico.com/story/2016/06/former-obama-insiders-seek-administrations-blessing-of-for-profit-college-takeover-224917; Peter Nicholas, "A Close-Knit Inner Circle," *Los Angeles Times*, November 9, 2008, http://articles.latimes.com/2008/nov/09/nation/na-obamacircle9; "Obama Kicks Back with 'FOB#1' (Friend of Barack)."

7 Shailagh Murray, "Obama's Circle, Chicago Remains the Tie that Binds," *Washington Post*, July 14, 2008, http://www.washingtonpost.com/wp-dyn/content/article/2008/07/13/AR2008071301904.html.

8 Harris, "Executive Profile: Martin Nesbitt, the First Friend."

9 Ibid.; Tom Henderson, "Forget those Chicago Rumors: Kip Kirkpatrick is Having Too Much Fun Building a Mortgage Giant to Go Anywhere," *Crain's Detroit Business*, February 22, 2013, http://www.crainsdetroit.com/article/20130222/BLOG007/130229940/forget-those-chicago-rumors.

10 Stratford and Hefling, "Bid to Buy For-Profit College Raises Questions"; *The Pritzker Traubert Family Foundation, Internal Revenue Service, Form 990*, 2014, http://990s.foundationcenter.org/990pf_pdf_archive/364/364347781/364347781_201412_990PF.pdf.

11 Mary Cass, "Illinois Fund Approves Equity Rebalancing, PE Commitment," *Money Management Letter*, September 1, 2015, ProQuest; "Hirings," *Pensions and Investments*, May 4, 2015, ProQuest.

12 "Private Equity with a Social Mission: After Running Parking Businesses, Martin Nesbitt, '89, Shifts Gears," *Chicago Booth Magazine*, accessed June 15, 2017, https://www.chicagobooth.edu/magazine/35/2/alumninews/alumni1.aspx.

13 Harris, "Executive Profile: Martin Nesbitt, the First Friend."

14 "Well-connected Chicago Emerging Manager," Partner Vine, accessed June 15, 2017, http://pe.partnervine.com/reviews/Vistria-Group.

15 "President Obama's Vacation in Hawaii," on Armchair Hawaii's website, December 2012, https://armchairhawaii.com/obama_hawaii/obama-hawaii-2012/; James Nye, "Obama begins Hawaiian Christmas vacation by playing golf with friend arrested for soliciting prostitution," *Daily Mail*, December 22, 2012, http://www.dailymail.co.uk/news/article-2252401/Obama-begins-Hawaiian-Christmas-vacation-playing-golf-friend-arrested-soliciting-prostitute.html.

16 "Basketball Superfan Obama Attends Syracuse-Marquette NCAA Game

with Friends After Playing First Round of Golf Since the Sequester," *Daily Mail*, March 30, 2013, http://www.dailymail.co.uk/news/article-2301719/Basketball-superfan-Obama-attends-Syracuse-Marquette-NCAA-game-friends-playing-round-golf-sequester.html.

17 Soyoung Kim, "Apollo Teams with Washington Insider for Education Deal: Sources," Reuters, January 12, 2016, http://www.reuters.com/article/us-apollo-education-m-a-apollo-global-idUSKCN0UQ23W20160112.

18 Associated Press, "Obama Tees Off Birthday Weekend with Round of Golf Ahead of Spending Time with Friends at Camp David," *New York Daily News*, August 3, 2013, http://www.nydailynews.com/news/politics/obama-tees-birthday-weekend-golf-article-1.1416782.

19 Tory Newmyer, "Exclusive: An Obama Friend Poaches a White House Aide," *Fortune*, September 2, 2014, http://fortune.com/2014/09/02/obama-friend-poaches-white-house-aide/.

20 "Jonathan Samuels," profile, LinkedIn, accessed October 19, 2017, https://www.linkedin.com/in/jonsamuels/.

21 Newmyer, "Exclusive: An Obama Friend Poaches a White House Aide."

22 Stratford and Hefling, "Bid to Buy For-Profit College Raises Questions."

23 Ashey A. Smith, "Reshaping the For-Profit," *Inside Higher Ed.*, July 15, 2015, https://www.insidehighered.com/news/2015/07/15/profit-industry-struggling-has-not-reached-end-road.

24 Goldie Blumenstyk, "Obama Singles Out For-Profit Colleges and Law Schools for Criticism," *Chronicle*, August 23, 2013, http://www.chronicle.com/article/Obama-Singles-Out-For-Profit/141253/.

25 Jon Marcus, "Can the University of Phoenix Rise from the Ashes?" http://time.com/money/4246709/can-the-university-of-phoenix-rise-from-the-ashes/.

26 Jared Borodinsky, "The Economic Dangers of For Profit Colleges," *The Higher Education Revolution*, December, 10, 2015, https://higheredrevolution.com/the-economic-dangers-of-for-profit-colleges-9b764b53518b.

27 Justin Hamilton, "Obama Administration Announces New Steps to Protect Students from Ineffective Career College Programs," U.S. Department of Education, press release, June 2, 2011, https://www.ed.gov/news/press-releases/obama-administration-announces-new-steps-protect-students-ineffective-career-college-programs.

28 "Final List of Private Meetings and Attendees Proposed Rule for Gainful Employment," U.S. Department of Education, December 15, 2010, https://www2.ed.gov/policy/highered/reg/hearulemaking/2009/privtemeetings2010.pdf.

29 "Short Seller Connected Senator Continues Error-Ridden Attack on Private Sector Schools," *PR Newswire*, September 22, 2011, https://www.thestreet.com/story/11257356/1/short-seller-connected-senator-continues-error-ridden-attack-on-private-sector-schools.html.

30 Citizens for Responsibility and Ethics in Washington Chief Counsel Anne L. Weismann to U.S. Department of Education Secretary Arne Duncan, March 11, 2011, http://edworkforce.house.gov/uploadedfiles/03.11.11_thompson_submit.pdf.

31 Ibid., Exhibit E.

32 Project on Government Oversight, "Did Education Department Officials Leak Market-Sensitive Info to Stock Traders?" press release, June 13, 2011, http://www.pogo.org/about/press-room/releases/2011/gc-ii-20110613.html.

33 Allie Grasgreen, "Obama Pushes For-Profit Colleges to the Brink," *Politico*, July 1, 2015, http://www.politico.com/story/2015/07/barack-obama-pushes-for-profit-colleges-to-the-brink-119613.

34 Ibid.

35 Angela Gonzales, "University of Phoenix Parent Under FTC Investigation, Stock Drops," *Phoenix Business Journal*, July 29, 2015, http://www.bizjournals.com/phoenix/news/2015/07/29/university-of-phoenix-parent-under-ftc.html.

36 John McCain, Senate Committee of Armed Services, Department of Defense: Actions Against the University of Phoenix Regarding the Voluntary Education Tuition Assistance Program (2016), https://www.republicreport.org/wp-content/uploads/2016/12/mccain-report-on-dod-actions-aganist-the-university-of-phoenix-12–16–16.pdf, 63.

37 David Halperin, "Top Democratic Lawyer Pushed Pentagon to End U. of Phoenix Suspension," *The Blog* (blog), *Huffington Post*, March 17, 2017, http://www.huffingtonpost.com/davidhalperin/top-democratic-lawyer-pus_b_9487600.html.

38 "For Profit College Chains Recruit Veterans to Keep GI Money Flowing," *Ames Tribune*, August 22, 2015.

39 Travis J. Tritten, "DoD Says 'Crappy' Process Led to University of Phoenix Probation," *Stars and Stripes*, November 29, 2016.

40 Stratford and Hefling, "Bid to Buy For-Profit College Raises Questions"; Matt Jarzemsky, Dana Cimilluca, and Austen Hufford, "Apollo Global Management in Talks to Buy Apollo Education," *Wall Street Journal*, January 11, 2016, https://www.wsj.com/articles/apollo-education-to-explore-strategic-alternatives-1452516388; Stock price graph for Apollo Education Group Inc, January 4, 2017–February 1, 2017, via Google Finance, accessed June 15, 2017, https://www.google.com/finance?cid=657521.

41 Halperin, "Top Democratic Lawyer Pushed Pentagon to End U. of Phoenix Suspension."

42 Lynn Sweet, "Obamas Host White House Dinner Saturday for Obama Foundation," *Chicago Sun-Times*, April 3, 2016, http://chicago.suntimes.com/politics/obamas-host-white-house-dinner-saturday-for-obama-foundation/; Shia Kapos, "At Obama's Vacation Spot, Politics Meet Golf,"

Taking Names (blog), August 9, 2016, http://shiakapos.com/at-obamas-vacation-spot-politics-meets-golf/; Mark Shanahan, "Vacationing Obama Back on a Vineyard Golf Course Monday," *Boston Globe*, August 15, 2016, https://www.bostonglobe.com/lifestyle/names/2016/08/15/vacationing-obama-back-vineyard-golf-course-monday/mkTreuQ0Hpr25vKbSx0M2H/story.html; Lynn Sweet, "Obama Fundraising in Chicago for Hillary, Duckworth, House Dems," *Chicago Sun-Times*, September 30, 2016, http://chicago.suntimes.com/politics/obama-pelosi-fundraise-for-house-democrats-in-chicago-oct-7/; Susan Crabtree, "Obama Sticks with Basketball on Election Day," *Washington Examiner*, November 8, 2016, http://www.washingtonexaminer.com/obama-sticks-with-basketball-on-election-day/article/2606775.

43 Ashey A. Smith, "Apollo Sale Approved, with Conditions," *Inside Higher Ed*, December 8, 2016, https://www.insidehighered.com/news/2016/12/08/education-department-approves-apollo-deal-conditions.

44 Josh Kosman, "Obama's Pal Catches Major Break in For-Profit College Deal," *New York Post*, January 3, 2017, http://nypost.com/2017/01/03/obamas-pal-catches-major-break-in-for-profit-college-deal/.

45 "Apollo Education Group, Inc. to Be Taken Private in $1.1 Billion Transaction," *Business Wire*, Feburary 8, 2016, http://www.businesswire.com/news/home/20160208005571/en/Apollo-Education-Group-Private-1.1-Billion-Transaction.

46 Maria Armental, "Apollo Education Shareholders Approve $1.14 Billion Buyout," *Wall Street Journal*, May 6, 2016, https://www.wsj.com/articles/apollo-education-shareholders-approve-1–14-billion-buyout-1462574658.

47 Russ Wiles, "University of Phoenix Parent Apollo Education Starts New Chapter as Private Firm," *AZ Central*, February 6, 2017, http://www.azcentral.com/story/money/business/economy/2017/02/06/university-phoenix-parent-apollo-education-starts-new-chapter-private-firm/97553824/.

48 Alexia Elejaide-Ruiz, "Arne Duncan: Police Can't Stop Chicago's Violence, Business Community Must Step Up," *Chicago Tribune*, October 15, 2016, http://www.chicagotribune.com/business/ct-arne-duncan-emerson-collective-1016-biz-20161014-story.html.

49 Jennifer Gould, Christina Pascucci, and Mary Beth McDade, "More than 35,000 Students Left Without Degrees After ITT Tech Announces Immediate Campus Closures," KTLA, September 6, 2016, http://ktla.com/2016/09/06/itt-tech-shuts-down-thousands-of-students-left-in-the-lurch/.

50 Kosman, "Obama's Pal Catches Major Break in For-Profit College Deal."

51 Stratford and Hefling, "Bid to Buy For-Profit College Raises Questions."

52 Ibid.

53 David J. Garrow, *Rising Star: The Making of Barack Obama* (New York: William Morrow, 2017), 463.

54 Wendy D. Fox, Ariel Investments, 2010, quarter 3, form 13-F (filed with

the Securities and Exchange Commission on November 12, 2010), https://www.sec.gov/Archives/edgar/data/936753/0000936753–10–000055.txt; "Ariel Appreciation Fund," *U.S. News and World Report*, accessed June 15, 2017, http://money.usnews.com/funds/mutual-funds/mid-cap-blend/ariel-appreciation-fund/caapx.

55 Wendy D. Fox, Ariel Investments, 2011, quarter 1, 13-F (filed with the Securities and Exchange Commission on May 13, 2011), https://www.sec.gov/Archives/edgar/data/936753/000093675311000054/hr13f1Q11.txt.

56 Ben Protess, "Grassley Questions Education Agency's Ties to Hedge Funds," *New York Times*, July 28, 2011, https://dealbook.nytimes.com/2011/07/28/grassley-questions-education-agencys-ties-to-wall-street/?mtrref=www.google.com&_r=0.

57 Noah Black, "Coalition Urges SEC to Investigate Possible Insider Trading by Short-Sellers Who Obtained Material Non-Public Information from U.S. Department of Education Officials in Violation of Department Ethics Rules," *Business Wire*, May 9, 2011, http://www.businesswire.com/news/home/20110509006236/en/Coalition-Urges-SEC-Investigate-Insider-Trading-Short-Sellers.

58 Juliet Eilperin, Lyndsey Layton, and Emma Brown, "U.S. Education Secretary Arne Duncan to Step Down at End of Year," *Washington Post*, October 2, 2015, https://www.washingtonpost.com/news/education/wp/2015/10/02/education-secretary-arne-duncan-reportedly-will-step-down-at-end-of-year/?utm_term=.77c1a64a6011; Ariel Investment Trust, 2017, N-1A (filed with the Securities and Exchange Commission on January 26, 2017), https://www.sec.gov/Archives/edgar/data/798365/0001193125 17019767/d282974d485bpos.htm.

59 Joel Poelhuis, "CFPB and 'the Great Unknown,'" *SNL Financial Services Daily*, October 14, 2011, ProQuest; Norbert Michel, "Dodd-Frank and the Consumer Financial Protection Bureau Put Squeeze on Private Payday Lenders," *Heritage Foundation*, November 4, 2015, http://www.heritage.org/markets-and-finance/report/dodd-frank-and-the-consumer-financial-protection-bureau-put-squeeze.

60 Fed. Deposit Insurance Corp., FDC 6494/03 (11–10), 2011 FDIC National Survey of Unbanked and Underbanked Households, Federal Deposit Insurance corporation (September 2012), https://www.fdic.gov/householdsurvey/2012_unbankedreport.pdf.

61 Robert DeYoung, Ronald J. Mann, Donald P. Morgan, and Michael R. Strain, "Reframing the Debate About Payday Lending," Federal Reserve Bank of New York, October 19, 2015, http://libertystreeteconomics.newyorkfed.org/2015/10/reframing-the-debate-about-payday-lending.html#.V02mGzUrJhE.

62 U.S. Department of Justice, "Justice Department and Consumer Financial Protection Bureau Pledge to Work together to Protect Consumers from

Credit Discrimination," press release, December 6, 2012, https://www.justice.gov/opa/pr/justice-department-and-consumer-financial-protection-bureau-pledge-work-together-protect.

63 First Cash Financial Services, FY 2012, form 10-K (annual report filed with the Securities and Exchange Commission in 2013), https://www.sec.gov/Archives/edgar/data/840489/000084048913000005/fcfs1231201210-k.htm.

64 Diane Katz, "The CFPB in Action: Consumer Bureau Harms Those It Claims to Protect," *Heritage Foundation*, January 22, 2013, http://www.heritage.org/housing/report/the-cfpb-action-consumer-bureau-harms-those-it-claims-protect.

65 Tim Zawacki, "CFPB Regulation Seen as Boon, Not Burden, to IPO-seeking Lender," *SNL Bank M&A Weekly*, May 7, 2012, ProQuest.

66 Staff of H.R. Comm. On Oversight and Government Reform, 113th Cong., Rep. on "The Department of Justice's 'Operation Choke Point': Illegally Choking Off Legitimate Businesses?" (2014), https://oversight.house.gov/wp-content/uploads/2014/05/Staff-Report-Operation-Choke-Point1.pdf.

67 Associated Press, "FDIC, Fed Must Face Payday Lenders' 'Choke Point' Lawsuit," *Bloomberg*, September 27, 2015, https://www.bloomberg.com/news/articles/2015–09–25/fdic-fed-must-face-payday-lenders-choke-point-lawsuit.

68 Cheryl Winokur Munk, "Building Fintech's Future: MagTek CEO Annmarie 'Mimi' Hart Is Eager to Witness the Potential of Futuristic Payment Technologies, from Wearables to 'Edibles' and 'Injectables.' But First, She Must Work with the Merchant Community to Address the Shortcomings of Today's Approaches to Payment Security," *ISO & Agent*, April 1, 2016, Nexis.

69 Tim Devaney, "Lawmakers Criticize Agencies over Payday Lending Regs," *The Hill*, April 8, 2014, http://thehill.com/regulation/finance/203003-lawmakers-criticize-agencies-over-payday-lending-regs.

70 Lydia DePillis, "Is the CFPB About to Break the Payday Lending Business Model?" *Washington Post*, March 25, 2014, https://www.washingtonpost.com/news/wonk/wp/2014/03/25/is-the-cfpb-about-to-break-the-payday-lending-business-model/?utm_term=.022ee442c84f.

71 "About ForwardLine: Meeting a Critical Need on Main Street," Forward Line, accessed June 15, 2017, https://www.forwardline.com/about-us. Note: these quotations were taken on 5/16/2017—as of 5/24/17, the language on the website had changed slightly.

72 "ForwardLine-CFO Placement (Interim to Perm)," Buxbaum HCS, December 11, 2015, http://www.buxbaumhcs.com/testimonials/forwardline-cfo-placement-interim-to-perm/.

73 Tim Zawacki, "Privately Held Payday Lender Gets Top Billing in Industry's Operation Choke Point Response," *SNL Financial Services Daily*, June 9, 2014, ProQuest.

74 "ForwardLine Consumer Complaints and Reviews," Consumer Affairs, last updated May 23, 2017, https://www.consumeraffairs.com/business-loans-and-financing/forwardline.html.

75 "ForwardLine Financial: Reviews," Best Company, accessed June 15, 2017, https://bestcompany.com/business-loans/company/forwardline/.

76 "The Economics and Regulation of the Freight Rail Industry," Fisher Colloquium, McDonough School of Business, Georgetown University, June 10, 2016; Norfolk Southern Corporation, FY 2012, form 10-K (annual report filed with the Securities and Exchange Commission in 2013), https://www.sec.gov/Archives/edgar/data/702165/000070216513000042/nsc12.htm: "The relaxation of economic regulation of railroads, following the Staggers Rail Act of 1980, included exemption from STB [Surface Transportation Board] regulation of the rates and most service terms for intermodal business (trailer-on-flat-car, container-on-flat-car), rail boxcar shipments, lumber, manufactured steel, automobiles, and certain bulk commodities such as sand, gravel, pulpwood, and wood chips for paper manufacturing. Further, all shipments that we have under contract are effectively removed from regulation for the duration of the contract. About 86% of our revenues comes from either exempt shipments or shipments moving under transportation contracts; the remainder comes from shipments moving under public tariff rates."

77 Pat Foran, ed., *Progressive Railroading Magazine* 56, no. 2 (February 2013), http://www.progressiverailroading.com/pr/graphics/13/02/pr0213.pdf.

78 Norfolk Southern Corporation, FY 2012, form 10-K (annual report filed with the Securities and Exchange Comission in 2013), https://www.sec.gov/Archives/edgar/data/702165/000070216513000042/nsc12.htm.

79 Bill Shuster, "Rail Re-Regulation May Be Catastrophic Public Policy," *The Hill*, March 11, 2010, http://thehill.com/special-reports/transportation-a-infrastructure-march-2010/86305-rail-re-regulation-may-be-catastrophic-public-policy-.

80 "About STB: Overview," Surface Transportation Board, accessed June 15, 2017, https://www.stb.gov/stb/about/overview.html; "About FRA," Federal Railroad Administration, accessed June 15, 2017, https://www.fra.dot.gov/Page/P0002.

81 Norfolk Southern, "Martin H. Nesbitt and John R. Thompson Elected to Norfolk Southern Board," news release, February 13, 2013, http://www.nscorp.com/content/nscorp/en/news/martin-h-nesbitt-and-john-r-thompson-elected-to-norfolk-southern-board.html.

82 Norfolk Southern Corporation, "Compensation of Directors," 2016 Proxy Statement, May 12, 2016, 19, http://www.nscorp.com/content/dam/nscorp/get-to-know-ns/investor-relations/proxy-statements/nsc_proxy_2016.pdf, 19.

83 Michelle Chan, "Politics in the American Airlines-U.S. Airways Merger and Antitrust Settlement," *Fordham Journal of Corporate and Financial Law* 20, no. 1 (2014): 175–201, http://ir.lawnet.fordham.edu/cgi/viewcontent.cgi?article=1400&context=jcfl.

84 Ibid.

85 Justin Elliott, "The American Way," *ProPublica*, October 11, 2016, https://www.propublica.org/article/airline-consolidation-democratic-lobbying-antitrust.

86 Andy Stonehouse, "Federal Pension Agency Reports Record Deficit," *Benefits Selling*, November 16, 2012, ProQuest; Jerry Geisel, "2012: Year in Review: Benefits Management," *Business Insurance*, December 24, 2012, ProQuest.

87 Andrew Harrer, "Five US Airlines Face Price-gouging Allegations," *CNBC*, July 24, 2015, http://www.cnbc.com/2015/07/24/five-us-airlines-face-price-gouging-allegations.html.

88 Linda Lloyd, "Airlines Ramp Up Effort to Renegotiate Open Skies Treaties with Gulf Carriers," *Philadelphia Inquirer*, March 6, 2015, Nexis.

89 Joan Lowy, "Big 3 Airlines Ask Government: Shield Us from Competition and Roll Back Consumer Protections," *U.S. News*, October 14, 2015, https://www.usnews.com/news/business/articles/2015/10/14/big-3-airlines-flexing-their-political-muscle-in-washington.

90 Michael A. Lindenberger, "American Takes Up Cause," *Dallas Morning News*, March 5, 2015, Nexis.

91 Bart Jansen, "Airline Execs Meet with Kerry over Gulf Carrier Subsidy Dispute," *USA Today*, September 18, 2015, https://www.usatoday.com/story/travel/flights/todayinthesky/2015/09/18/airline-subsidies-secretary-state-john-jerry-american-delta-united/72392048/.

92 Chico Harlan, "Landing a Mega-Merger: The Last Days of US Airways," *Washington Post*, September 25, 2015, https://www.washingtonpost.com/business/the-last-days-of-us-airways/2015/09/25/f5530686–60a6–11e5–8e9e-dce8a2a2a679_story.html?utm_term=.bded89df38b4; Gregory Karp, "Martin Nesbitt Named to American Airlines Board," *Chicago Tribune*, November 12, 2015, http://www.chicagotribune.com/business/ct-martin-nesbitt-american-airlines-1113-biz-20151111-story.html.

93 Robert Fulford, "Closing Down the Daschle Racket," *National Post*, February 7, 2009, ProQuest.

94 Edward-Isaac Dovere, "The Man Building Barack Obama's Future," *Politico*, August 20, 2015, https://www.politico.com/story/2015/08/marty-nesbitt-barack-obama-foundation-chair-121544.

CHAPTER 10: MORE SMASHING AND GRABBING

1 "Obama on Building Coal Plants in the United States," YouTube video, 00:12, remarks by Barack Obama from an interview with *San Francisco Chronicle*, posted by "CoalMatters08," November 2, 2008, https://www.youtube.com/watch?v=4aTf5gjvNvo; Michael Bastasch, "Flashback 2008: Obama Promised to 'Bankrupt' Coal Companies," *Daily Caller*, August 3, 2015, http://dailycaller.com/2015/08/03/flashback-2008-obama-promised-to-bankrupt-coal-companies/.

2 Senator Barack Obama, "Real Leadership for a Clean Energy Future," (speech, Portsmouth, NH, October 8, 2007), *Grist*, http://grist.org/article/obamas-speech/.

3 Richard L. Revesz and Jack Lienke, "How Obama Went from Coal's Top Cheerleader to Its No. 1 Enemy," *Grist*, February 15, 2016, http://grist.org/climate-energy/how-obama-went-from-coals-top-cheerleader-to-its-no-1-enemy/.

4 Ibid.

5 President Barack Obama, "Remarks by the President on Clean Energy" (speech, Newton IA, April 22, 2009), *CBS News*, http://www.cbsnews.com/news/transcript-obamas-earth-day-speech/.

6 Andy Rowell, "Obama Targets $30 Billion Oil and Gas Subsidies," *Oil Change International*, February 27, 2009, http://priceofoil.org/2009/02/27/obama-targets-30-billion-oil-and-gas-subsidies/.

7 Peter Waldman, "Exxon vs. Obama," *Entrepreneur*, March 30, 2009, https://www.entrepreneur.com/article/200986.

8 Campbell Robertson, "Search Continues After Oil Rig Blast," *New York Times*, April 21, 2010, http://www.nytimes.com/2010/04/22/us/22rig.html.

9 Nicholas Johnston and Hans Nichols, "Obama Says New Oil Leases Must Have More Safeguards," *Bloomberg*, May 1, 2010, https://www.bloomberg.com/news/articles/2010–04–30/new-offshore-oil-drilling-must-have-safeguards-obama-says; H. Josef Hebert and Julie Pace, "Obama Says No New US Offshore Drilling Leases Until New Safeguards in Place," *Canadian Press,* April 30, 2010, Ebsco; "Oil Spill Reaches Mississippi River," *CBS News*, April 29, 2010, http://www.cbsnews.com/news/oil-spill-reaches-mississippi-river/.

10 Bruce Alpert and Rebecca Mowbray, "President Barack Obama Suspends Drilling at 33 Wells in the Gulf of Mexico," *Times-Picayune*, May 27, 2010, http://www.nola.com/news/gulf-oil-spill/index.ssf/2010/05/president_barack_obama_suspend.html.

11 Louis Jacobson, "Obama Blames MMS for Being Captive to Oil Industry," *Politifact,* June 17, 2010, http://www.politifact.com/truth-o-meter/statements/2010/jun/17/barack-obama/obama-blames-mms-being-captive-oil-industry/.

12 American Legislative Exchange Council, *EPA's Regulatory Train Wreck: Strategies for State Legislators* (Washington, DC: ALEC, 2011), https://www3.epa.gov/region1/npdes/merrimackstation/pdfs/ar/AR-1179.pdf.

13 H.R. Committee On Natural Resources, 112 Cong., "Reversing President Obama's Offshore Moratorium Act," H.R. Rep. No. 112–69 (2011), https://www.gpo.gov/fdsys/pkg/CRPT-112hrpt69/html/CRPT-112hrpt69.htm.

14 H.R. Committee On Natural Resources, "Obama Administration Imposes Five-Year Drilling Ban on Majority of Offshore Areas," press release, No-

vember 8, 2011, http://naturalresources.house.gov/newsroom/documents ingle.aspx?DocumentID=267985.

15 John M. Broder, "Obama's Bid to End Oil Subsidies Revives Debate," *New York Times*, January 31, 2011, http://www.nytimes.com/2011/02/01/science/earth/01subsidy.html.

16 Patrick Reis, "W. Va. Sues Obama, EPA Over Mining Coal Regulations," *New York Times*, October 6, 2010, http://www.nytimes.com/gwire/2010/10/06/06greenwire-wva-sues-obama-epa-over-mining-coal-regulation-48964.html.

17 Tom Osborn, "Trade of the Week: Obama Scuttles Coal Stocks," Dow Jones, November 8, 2012; Associated Press, "Election Over, U.S. Stocks Drop on Challenges to Come," *CBS News*, November 7, 2012, http://www.cbsnews.com/news/election-over-us-stocks-drop-on-challenges-to-come/.

18 American Legislative Exchange Council, *EPA's Regulatory Train Wreck*; David Gardner, "Republican Tsunami: Democrats Lose Control of the House as Voters Slam Obama with Worst Losses for 62 Years," *Daily Mail*, November 3, 2010, http://www.dailymail.co.uk/news/article-1326053/MID-TERM-ELECTIONS-2010-Democrats-lose-House-Republican-tsunami.html.

19 Robert Rapier, "The Energy Sector Under President Obama: The Picture Is Sunnier Than You Think," *Forbes*, December 18, 2012, https://www.forbes.com/sites/energysource/2012/12/18/the-energy-sector-under-president-obama-ignore-the-drama-big-oil-will-do-just-fine/#424343992ee2.

20 "As He Leaves Office, Obama Makes A Last-Ditch Effort to Kill Off Fossil Fuels," *Investors Business Daily*, December 20, 2016, http://www.investors.com/politics/editorials/as-he-leaves-office-obama-makes-a-last-ditch-effort-to-kill-off-fossil-fuels/.

21 Richard L. Revesz and Jack Lienke, *Struggling for Air: Power Plants and the "War on Coal"* (New York: Oxford University Press, 2016), 157.

22 Matt Egan, "Coal Companies Have Been Scorched Under Obama," *CNN Money*, August 3, 2015, http://money.cnn.com/2015/08/03/investing/coal-obama-climate-change/.

23 H.R. Comm. On Natural Resources, "Obama Administration Imposes Five-Year Drilling Ban on Majority of Offshore Areas," press release, November 8, 2011, https://naturalresources.house.gov/newsroom/documentsingle.aspx?DocumentID=267985; Wendy D. Fox, Ariel Investments, 2010, quarter 3, form 13-F (filed with the Securities and Exchange Commission on November 12, 2010), https://www.sec.gov/Archives/edgar/data/936753/000093675310000055/0000936753–10–000055.txt; "Obama Says No New US Offshore Drilling Leases Until New Safeguards in Place," Canadian Press, April 30, 2010, Ebsco; "Gulf Island Fabrication Inc.," historical quote, Market Watch, accessed January 8, 2018,

https://www.marketwatch.com/investing/stock/GIFI/historical?siteid =mktw&date=04%2F%2F29%2F2010&x=0&y=0; Wendy D. Fox, Ariel Investments, 2010, quarter 2, form 13-F (filed with the Securities and Exchange Commission on August 13, 2010), https://www.sec.gov/Archives /edgar/data/936753/000093675310000049/0000936753–10–000049.txt; Wendy D. Fox, Ariel Investments, form 13-F (filed with the Securities and Exchange Commission on November 12, 2010), https://www.sec .gov/Archives/edgar/data/936753/000093675310000055/0000936753 –10–000055.txt.; Dianne Tordillo, "Ariel Fund's John Rogers Stocks Up on Contango Oil," Guru Focus, December 6, 2012, https://www.gurufocus .com/news/200614/ariel-funds-john-rogers-stocks-up-on-contango-oil; "Contango Oil & Gas Company (MCF)," Yahoo Finance, accessed January 8, 2018, https://finance.yahoo.com/quote/MCF/history?period1 =1270094400&period2=1280635200&interval=1d&filter=history &frequency=1d; Wendy D. Fox, Ariel Investments, 2011, quarter 2, form 13-F (filed with the Securities and Exchange Commission on August 12, 2011), https://www.sec.gov/Archives/edgar/data/936753/0000936753 11000062/0000936753–11–000062.txt/; Wendy D. Fox, Ariel Investments, 2013, quarter 4, form 13-F (filed with the Securities and Exchange Commission on February 14, 2013), https://www.sec.gov /Archives/edgar/data/936753/000093675313000051/0000936753–13 –000051.txt.

24 Wendy D. Fox, Ariel Investments, 2011, quarter 1, form 13-F (filed with the Securities and Exchange Commission on May 13, 2011), https://www .sec.gov/Archives/edgar/data/936753/000093675311000054/0000936753 –11–000054.txt; Wendy D. Fox, Ariel Investments, 2011, quarter 2, form 13-F (filed with the Securities and Exchange Commission on August 12, 2011), https://www.sec.gov/Archives/edgar/data/936753/00009367 5311000062/0000936753–11–000062.txt; United States Securities and Exchange Commission, Form 13F Information Table, accessed January 8, 2018, https://www.sec.gov/Archives/edgar/data/936753/000093675313000119 /xslForm13F_X01/Form13FTableExport.xml.

25 "John W. Rogers," White House Visitor Logs, InsideGov, accessed January 8, 2018, http://white-house-logs.insidegov.com/l/84383517/John-W -Rogers#John%20W.%20Rogers%27s%20Visits&s=4mb6i3

26 "John W. Rogers," http://white-house-logs.insidegov.com/l/71191229 /John-W-Rogers; "Senior Advisor Brian Deese," White House Archives, accessed January 8, 2018, https://obamawhitehouse.archives.gov /administration/senior-leadership/brian-deese.

27 Evan Lehmann, "Emanuel's Replacement Might Calm the Climate Debate," *New York Times*, October 1, 2010, http://www.nytimes.com /cwire/2010/10/01/01climatewire-emanuels-replacement-might-calm -the-climate-75175.html.

28 Frank Verrastro, "The Evolution of US Energy Policy: 2008–2011," Center for Strategic and International Studies: Energy & Natural Security Program, accessed June 16, 2017, http://www.ncac-usaee.org/pdfs/2011_02Verrastro .pdf.

29 "Company Board of Directors," Ariel Investments, accessed June 16, 2017, https://www.arielinvestments.com/board-of-directors/.

30 Amy Tikkanen, Encyclopaedia Britannica Online, s.v. "Steyer, Tom," accessed June 16, 2017, https://www.britannica.com/biography/Tom-Steyer; Ryan Lizza, "The President and the Pipeline," *New Yorker*, September 16, 2013, http://www .newyorker.com/magazine/2013/09/16/the-president-and-the-pipeline.

31 Lizza, "The President and the Pipeline."

32 Richard Valdmanis, Fergus Jensen, and Sonali Paul, "From Black to Green: U.S. Billionaire's 'Road to Damascus,'" Reuters, May 13, 2014, http://www .reuters.com/article/us-usa-steyer-coal-insight-idUSBREA4C06B20140513.

33 David Callahan, "Traitors to Their Class," *New Republic*, June 25, 2010, https://newrepublic.com/article/75615/traitors-their-class.

34 Tom Steyer, interview by Nick Stockton, "The Billionaire on a Mission to Save the Planet from Trump," *Wired*, March 23, 2017, https://www.wired .com/2017/03/tom-steyer-interview/.

35 "Tom Steyer Shows 9.9% Stake in FreightCar America (RAIL)," *Market-Folly.com*, January 20, 2010, http://www.marketfolly.com/2010/01/thomas -steyer-shows-99-stake-in.html.

36 "The Top 15 Stocks Farallon Capital Management Is Buying," Seeking Alpha, May 18, 2011, https://seekingalpha.com/article/270579-the-top-15 -stocks-farallon-capital-management-is-buying.

37 Sabrina Tavernise, "Report Faults Mine Owner for Explosion That Killed 29," *New York Times*, May 19, 2011, http://www.nytimes.com/2011/05/20 /us/20mine.html.

38 Megan Barnett, "A New Roadmap for Coal," *Fortune*, December 20, 2011, http://fortune.com/2011/12/20/a-new-roadmap-for-coal/.

39 PR Newswire, "Alpha Natural Resources Acquires Massey Energy Company, Creating a Global Leader in Metallurgical Coal Supply," news release, June 1, 2011, http://www.prnewswire.com/news-releases/alpha-natural -resources-acquires-massey-energy-company-creating-a-global-leader-in -metallurgical-coal-supply-122946268.html.

40 Tom Steyer, "DNC Speech" (speech, Democratic National Committee, Charlotte, NC, September 5, 2012), *Politico*, http://www.politico.com /news/stories
/0912/80762_Page2.html.

41 Aaron Wormus, "Do Hedge Fund Managers Really Love Mitt Romney and Hate Obama?" *HedgeCo.net*, October 1, 2012, http://www.hedgeco.net/news /10/2012/do-hedge-fund-managers-really-like-mitt-romney-and-hate-obama .html.

42 Tom Steyer and John Podesta, "We Don't Need More Foreign Oil and Gas," *Wall Street Journal*, January 24, 2012, https://www.wsj.com/articles/SB10001424052970203718504577178872638705902?mg=id-wsj.

43 Juliet Eilperin and Steven Mufson, "Obama Administration Rejects Keystone XL Pipeline," *Washington Post*, January 18, 2012, https://www.washingtonpost.com/national/health-science/obama-administration-to-reject-keystone-pipeline/2012/01/18/gIQAPuPF8P_story.html?utm_term=.ff9027804a3d.

44 Lizza, "The President and the Pipeline."

45 "Farallon Capital," *Hedge Fund Wisdom*, May 2012, 60, http://www.hedgefundwisdom.com/wp-content/uploads/2012/08/Free_Q1_2012_Issue.pdf.

46 Maria Gallucci, "California Awaits Tar Sands Legal Ruling," *Guardian*, January 20, 2012, https://www.theguardian.com/environment/2012/jan/20/canada-tar-sands-legal-ruling.

47 Lizza, "The President and the Pipeline."

48 Ibid.

49 Mark C. Wehrly, Farallon Capital Management, 2012, form 13-F, quarter 1 (filed with the Securities and Exchange Commission on May 15, 2012), https://www.sec.gov/Archives/edgar/data/909661/000142210712000059/0001422107–12–000059.txt.

50 "Who We Are: Principals," on Internet Archive, accessed June 5, 2017, https://web.archive.org/web/20120221192440/; http://www.faralloncapital.com:80/farallon/principals_thomas_steyer.htm.

51 Kerry A. Dolan, "California Hedge Fund Billionare Tom Steyer To Step Down, Focus on Philanthropy," *Forbes*, October 23, 2012, https://www.forbes.com/sites/kerryadolan/2012/10/23/california-hedge-fund-billionaire-tom-steyer-to-step-down-at-farallon/#f6e495c7ad37.

52 Lizza, "The President and the Pipeline."

53 Garrow, *Rising Star*, 892.

54 "George Soros Interview 60 Minutes [FULL]," YouTube video, 13:27, George Soros, interview with Steve Kroft, 60 Minutes, CBS, December 20, 1998, posted by "ReasonReport," November 12, 2016, https://www.youtube.com/watch?v=Ar9EfzSnMZc.

55 Richard W. Rahn, "You Lose, Soros Wins," *Washington Times*, October 24, 2008, https://www.cato.org/publications/commentary/you-lose-soros-wins.

56 Nicola Clark, "Soros Loses Challenge to Insider Trading Conviction," *New York Times*, October 6, 2011, https://dealbook.nytimes.com/2011/10/06/soros-loses-challenge-to-insider-trading-conviction/?_r=0.

57 Beth Graddon-Hodgson, "George Soros Pledges $1.1 Billion to Fund Climate Change Initiatives," *Clean Technica*, October 14, 2009, https://cleantechnica.com/2009/10/14/george-soros-pledges-11-billion-to-fund-climate-change-initiatives/.

58 "Who Killed the Vote on Fracking?" *Boulder Weekly*, October 2, 2014, http://www.boulderweekly.com/news/who-killed-the-vote-on-fracking mdashglossary/; Mark Svenvold, "Cleantech Hits the Jackpot: George Soros to Invest $1 Billion in Green Energy," *AOL Finance*, October 12, 2009, https://www.aol.com/article/2009/10/12/cleantech-hits-the-jackpot-george-soros-to-invest-1-billion-in/19193275/.

59 Gregory Zuckerman, "Soros, Silver Lake Make Clean-Energy Bet," *Wall Street Journal*, February 24, 2011, https://www.wsj.com/articles/SB1000142405274870452050457616278125478416 2; Timothy P. Carney, "Democratic Donors Benefit from Democratic Policies," *Washington Examiner*, October 17, 2014, http://www.washingtonexaminer.com/democratic-donors-benefit-from-democratic-policies/article/2554908.

60 "George Soros Quantum Fund Investment Portfolio as of Q1/2012," Investing.com, June 7, 2012, https://www.investing.com/analysis/george-soros-quantum-fund-investment-portfolio-as-of-q1–2012–125778.

61 Linda Formella, "U.S. Exporters to Benefit from a $2 Billion Ex-Im Bank Preliminary Commitment to Brazil's Petrobras," U.S. Export-Import Bank, news release, May 6, 2009, http://www.exim.gov/news/us-exporters-benefit-2-billion-ex-im-bank-preliminary-committment-brazils-petrobras; "George Soros Top Stocks," *Guru Focus*, December 7, 2009, http://www.gurufocus.com/news/78292.

62 "A Bill to Make Soros Richer," *Investor's Business Daily*, June 14, 2011, http://www.investors.com/politics/editorials/a-bill-to-make-soros-richer/.

63 Katya Wachtel, "Here's Who Made a Fortune on Today's Huge Coal Deal," *Businessinsider.com*, May 2, 2011, http://www.businessinsider.com/heres-who-made-bank-today-on-the-arch-coal-buyout-of-international-coal-2011-5.

64 "George Soros Raises Gold ETF Holdings by Half, Nearly Triples Freeport Stake," *Mining News Magazine*, November 16, 2012, http://www.miningnewsmagazine.org/?p=349.

65 Malia Zimmerman, "Billionaire George Soros Warms Up to Coal as Stock Prices Hit Bottom," *Foxnews.com*, August 19, 2015, http://www.foxnews.com/us/2015/08/19/billionaire-george-soros-warms-up-to-coal-as-stock-prices-hit-bottom.html.

66 Thomas Landstreet, "Soros Doesn't Like Coal Stocks; He Likes Money," *Forbes*, August 28, 2015, https://www.forbes.com/sites/thomaslandstreet/2015/08/28/soros-doesnt-like-coal-stocks-he-likes-money/#324c07d91f4d.

67 Edward Vranic, "Why Would George Soros and Leon Black Buy Arch Coal?" Seeking Alpha, September 10, 2015, https://seekingalpha.com/article/3502906-george-soros-leon-black-buy-arch-coal.

68 Zimmerman, "Billionaire George Soros Warms Up to Coal as Stock Prices Hit Bottom."

69 Chris Francescani, "Paul Soros, Shipping Titan and Older Brother to George Soros, Dies at 87," ed. Alex Dobuzinskis and Bill Trott, Reuters, June 16,

2013, http://www.reuters.com/article/us-usa-paulsoros-idUSBRE95F01I20130616.

70 Timothy Cama, "Coal Plant Shutdowns Predicted to Double Under EPA Climate Rule," *The Hill*, May 22, 2015, http://thehill.com/policy/energy-environment/242931-study-coal-plant-shutdowns-would-more-than-double-under-epa-climate; James E. McCarthy, Cong. Research Serv., R43127, "EPA Standards for Greenhouse Gas Emissions from Power Plants: Many Questions, Some Answers (2013)," https://fas.org/sgp/crs/misc/R43127.pdf.

71 Chris Mooney, "Study: Coal Industry Lost Nearly 50,000 Jobs in Just Five Years," *Washington Post*, April 1, 2015, https://www.washingtonpost.com/news/energy-environment/wp/2015/04/01/the-decline-in-coal-jobs-in-one-chart/?utm_term=.0f44a2248c6c.

72 Dan Lowrey, "George Soros Discloses New Stakes in Coal Producers Peabody, Arch," S&P Global, August 14, 2015, https://www.snl.com/interactiveX/Article.aspx?cdid=A-33567529–11817&FreeAccess=1.

73 "Dow Jones U.S. Coal Index," prices for period between February 8, 2010–February 6, 2017, *Google Finance*, accessed June 21, 2017 https://www.google.com/finance/historical?cid=4931635&startdate=Feb+8%2C+2010&enddate=Feb+6%2C+2017&num=30&ei=rduYWOCGHc6xmAHCjpHIDQ.

74 "David E. Shaw," DE Shaw and Co., accessed June 21, 2017, https://www.deshaw.com/Founder.shtml.

75 Open Secrets, "Barrack Obama's Bundlers," 2012 Presidential Race, accessed June 21, 2017, https://www.opensecrets.org/pres12/bundlers.php.

76 Sarah Bryner, et al., "Organizing for Action: Who's giving to Obama-Linked Nonprofit?" *Center for Responsive Politics*, June 17, 2014, https://www.opensecrets.org/news/2014/06/organizing-for-action-whos-giving-to-obama-linked-nonprofit/; "Contributors," on Barack Obama's website, accessed June 21, 2017, https://www.obama.org/contributors/.

77 American Academy of Arts and Sciences, "President Obama Appoints Fellows of the Academy to Science and Technology Advisory Council," press release, April 29, 2009, https://www.amacad.org/content/news/pressReleases.aspx?pr=79.

78 Andrew C. Revkin, "Coal Trends Still Rule Climate Talks," *Dot Earth* (blog), *New York Times*, November 29, 2010, https://dotearth.blogs.nytimes.com/2010/11/29/coal-trends-still-rule-climate-talks/.

79 Nathaniel Heller and Derrick Wetherell, "Democratic Fundraiser, Cyber-Investor Enjoys Access to White House, Gore," *Center for Public Integrity*, August 17, 2000, https://www.publicintegrity.org/2000/08/17/3274/democratic-fundraiser-cyber-investor-enjoys-access-white-house-gore; American Academy of Arts and Sciences, "President Obama Appoints Fellows of the Academy to Science and Technology Advisory Council."

80 *25th Annual Global Power Markets Conference*, "Brave New World: Preparing for a Post-Carbon Future," April 11–13, 2010, https://www.platts.com/IM.Platts.Content/ProductsServices/ConferenceandEvents/2010/pc012/presentations/Peter_Maloney-intro.pdf?S=n.

81 First Wind Holdings, 2010, form S-1 (filed with the Securities and Exchange Commission on October 25, 2010), https://www.sec.gov/Archives/edgar/data/1434804/000104746910008828/a2195887zs-1a.htm; Joe Weisenthal, "At D.E. Shaw, Larry Summers Worked Just One Day A Week," *BusinessInsider.com*, April 6, 2009, http://www.businessinsider.com/at-de-shaw-larry-summers-worked-just-one-day-a-week-2009–4.

82 Stephen Taub, "Morning Brief: Trian Raises Money for New SPV," *AR: Absolute Return + Alpha*, August 21, 2015, ProQuest.

83 President Barack Obama, "Remarks by the President on Climate Change," (speech, Washington, DC, June 15, 2013), https://obamawhitehouse.archives.gov/the-press-office/2013/06/25/remarks-president-climate-change.

84 "Cheniere LNG Corporate Summary," company report, on Wood Mackenzie's website, July 2016, https://www.woodmac.com/reports/lng-cheniere-lng-corporate-summary-40090237.

85 Ben Lefebvre, "Cheniere: First LNG Export Facility to Start in Late 2015," *Wall Street Journal*, February 22, 2013, https://www.wsj.com/news/articles/SB10001424127887323549204578320051361543838; "FERC Process," Cheniere, accessed June 22, 2017, http://www.cheniere.com/terminals/sabine-pass/trains-1–6/ferc-process/.

86 Patrick Rucker and Jeff Mason, "Heather Zichal Resigns: Obama's Top Climate Adviser to Step Down After 5 Years at White House," *Huffington Post*, December 7, 2013, http://www.huffingtonpost.com/2013/10/07/heather-zichal-obama-adviser_n_4059066.html.

87 Steve Horn, "Heather Zichal, Former Obama Energy Aide, Named to Board of Fracked Gas Exports Giant Cheniere," *Desmog* (blog), June 20, 2014, https://www.desmogblog.com/2014/06/20/heather-zichal-former-obama-energy-aide-named-board-member-lng-exports-giant-cheniere; "Board of Directors," on Cheniere's website, accessed June 22, 2017, http://www.cheniere.com/about-us/cheniere-energy/board-of-directors/.

CHAPTER 11: A REAL ESTATE MOGUL GOES TO WASHINGTON

1 Charles V. Bagli, "Trump Sells Hyatt Share to Pritzkers," *New York Times*, October 8, 1996, http://www.nytimes.com/1996/10/08/business/trump-sells-hyatt-share-to-pritzkers.html?mcubz=2.

2 Nina Easton, "The Fascinating Life of Penny Pritzker (So far)," *Fortune*, July 2, 2014, http://fortune.com/2014/06/02/fortune-500-pritzker/; Janet Novack, "5 Questions Congress Should Ask Obama Commerce Secretary

Nominee Penny Pritzker," *Forbes*, May 18, 2013, https://www.forbes.com/sites/janetnovack/2013/05/18/5-questions-congress-should-ask-obama-commerce-nominee-penny-pritzker/#23f7ed6824e7.

3 John Fund, "Commerce Secretary Penny Pritzker?" *National Review*, February 8, 2013, http://www.nationalreview.com/article/340157/commerce-secretary-penny-pritzker-john-fund.

4 Gus Russo, *Supermob: How Sidney Korshak and His Criminal Associates Became America's Hidden Power Brokers* (New York: Bloomsbury, 2006), 68, 137.

5 Nancy Rivera Brooks, "Rooms with a View: Chance Encounter Led to Creation of Rapidly Expanding Hyatt Hotels Chain," *Los Angeles Times*, November 24, 1987, http://articles.latimes.com/1987–11–24/business/fi-24332_1_pritzker-family.

6 Paul O'Donnell, "What Made Me: Secretary of Commerce Penny Pritzker," *Washingtonian*, Feburary 6, 2014, http://www.washingtonian.com/2014/02/06/what-made-me-secretary-of-commerce-penny-pritzker/.

7 "Penny Pritzker: Founder and Chairman," on PSP Capital Partners' website, accessed July 13, 2017, https://www.pspcapital.com/penny-pritzker/.

8 Adam Clark Estes, "Obama's Apparent Pick for Commerce Secretary Has a History of Shady Commerce," *Atlantic*, February 6, 2013, https://www.theatlantic.com/politics/archive/2013/02/obamas-apparent-pick-commerce-secretary-has-history-shady-commerce/318588/.

9 Easton, "Fascinating Life of Penny Pritzker"; "Media Inquiries," University of Chicago Law School, accessed July 13, 2017, http://www.law.uchicago.edu/media; "Executive Profile: Martin Nesbitt, the First Friend," *Chicago Tribune*, January 21, 2013, http://www.chicagotribune.com/business/ct-biz-0121-executive-profile-nesbitt-20130121-story.html.

10 Garrow, *Rising Star*, 1021.

11 Viveca Novak, "Pritzker and Froman, by the Numbers," *Opensecrets.org*, May 2, 2013, https://www.opensecrets.org/news/2013/05/pritzker-and-froman-by-the-numbers/.

12 Jodi Kantor and Nicholas Confessore, "Leading Role in Obama '08, but Backstage in '12," *New York Times*, July 15, 2012, http://www.nytimes.com/2012/07/15/us/politics/penny-pritzker-had-big-role-in-obama-08-but-is-backstage-in-12.html; Kori Rumore and Ryan Marx, "In Obama's Words: 20 Memorable Quotes from Chicago Appearances," *Chicago Tribune*, January 11, 2017, http://www.chicagotribune.com/news/ct-barack-obama-chicago-speeches-htmlstory.html; Novak, "Pritzker and Froman, by the Numbers."

13 Edward McClelland, "Obama's Sugar Mama," *NBC Chicago*, July 16, 2012, http://www.nbcchicago.com/blogs/ward-room/Obamas-Sugar-Mama-162611046.html.

14 Easton, "Fascinating Life of Penny Pritzker."

15 Kantor and Confessore, "Leading Role in Obama '08, but Backstage in '12."

16 Novack, "5 Questions Congress Should Ask Obama Commerce Nominee Penny Pritzker."
17 Sara Jaffe, "Why 'Unions Don't Like' Penny Pritzker," *Aljazeera*, May 8, 2013, http://www.aljazeera.com/indepth/opinion/2013/05/20135873932733462.html.
18 Easton, "Fascinating Life of Penny Pritzker."
19 Mike Allen, "Pritzker Turns Down Commerce," *Politico*, November 20, 2008, http://www.politico.com/story/2008/11/pritzker-turns-down-commerce-015827.
20 American Presidency Project, "President-Elect Barack Obama Establishes President's Economic Recovery Advisory Board," press release, November 26, 2008, http://www.presidency.ucsb.edu/ws/index.php?pid=84977; "Member Bios: Penny Pritzker," President's Economic Recovery Advisory Board, accessed July 14, 2017, https://obamawhitehouse.archives.gov/administration/eop/perab/members/pritzker.
21 "Member Bios: Penny Pritzker," President's Council on Jobs and Competitiveness, accessed July 14, 2017, https://obamawhitehouse.archives.gov/administration/advisory-boards/jobs-council/members/pritzker.
22 Kantor and Confessore, "Leading Role in Obama '08, but Backstage in '12."
23 "History & Mission," on Artemis Real Estate Partners' website, accessed July 14, 2017, http://www.artemisrep.com/history-mission/.
24 "Illinois Municipal Sets $283 Million for Real Estate," *Pensions and Investments*, November 20, 2015, http://www.pionline.com/article/20151120/ONLINE/151129991/illinois-municipal-sets-283-million-for-real-estate; Artemis Real Estate Partners, Institutional Real Estate Letter-Americas, 2013, http://www.artemisrep.com/media/34247/institutional-real-estate-sponsor-profile-artemis-real-estate-partners.pdf.
25 Katherine Skiba, "Pritzker Opens the Books on Finances; Obama Nominee Discloses Wealth, Corporate Ties," *Chicago Tribune,* May 16, 2013.
26 Brian Wingfield, "Pritzker Understated Income, Files Amended Disclosure," *Bloomberg*, May 22, 2013, https://www.bloomberg.com/news/2013–05–22/pritzker-understated-income-files-amended-disclosure.html/.
27 Veronique de Rugy, "Department of Cronyism: The Department of Commerce's Economic Development Administration is a Favor-dispensing Machine," *Reason*, May 2014, http://reason.com/archives/2014/04/16/department-of-cronyism.
28 Letter from Penny Pritzker, CEO of PSP Capital Partners and Pritzker Realty Group, to Barbara Fredericks, Assistant General Counsel for Administration, United States Department of Commerce, May 8, 2013 (on file with Document Cloud), https://assets.documentcloud.org/documents/701133/penny-s-pritzker-ea.txt.

29 Advanced Data Search: Hyatt Hotel Contracts, Grants, Loans, Other Financial Assistance, for the Fiscal Years 2008–2016, *USASpending.gov* webpage, accessed July 14, 2017, https://www.usaspending.gov/Pages/Advanced Search.aspx?k=%22Hyatt%20hotel%22.

30 Letter from Penny Pritzker to Barbara Fredericks.

31 Melissa Harris, "A Day with Commerce Secretary Penny Pritzker," *Chicago Tribune*, December 5, 2014, http://www.chicagotribune.com/business /ct-confidential-penny-pritzker-1207-biz-20141205-column.html.

32 Hyatt, "Hyatt Announces Plans for a Second Hyatt Regency Hotel in Greater Moscow," news release, July 9, 2013, http://newsroom.hyatt .com/2013–07–09-HYATT-ANNOUNCES-PLANS-FOR-A-SECOND -HYATT-REGENCY-HOTEL-IN-GREATER-MOSCOW.

33 Hyatt, "Hyatt Announces Plans for Three New Hotels in the Kingdom of Saudi Arabia; Agreements for Park Hyatt Riyadh, Grand Hyatt Jeddah, and Hyatt Regency Jeddah Bring Total to Six Hotels Under Development in Saudi Arabia," news release, February 2, 2012, http://newsroom.hyatt .com/2012–02–02-HYATT-ANNOUNCES-PLANS-FOR-THREE -NEW-HOTELS-IN-THE-KINGDOM-OF-SAUDI-ARABIA -AGREEMENTS-FOR-PARK-HYATT-RIYADH-GRAND -HYATT-JEDDAH-AND-HYATT-REGENCY-JEDDAH-BRING -TOTAL-TO-SIX-HOTELS-UNDER-DEVELOPMENT-IN -SAUDI-ARABIA.

34 Burgess Everett, "On Pritzker, Where's the Outrage?" *Politico*, June 25, 2013, http://www.politico.com/story/2013/06/penny-pritzker-commerce -secretary-093321; Elspeth Reeve, "Senate Confirms Penny Pritzker to Be Commerce Secrtary, 91 to 1," *Atlantic*, June 25, 2013, https://www .theatlantic.com/politics/archive/2013/06/senate-confirms-penny -pritzker-be-commerce-secretary-97–1/313919/.

35 "Federal Real Property: Strategy Needed to Address Agencies' Long-Standing Reliance on Costly Leasing," GAO-08–197, February 25, 2008; Chuck Neubauer and Sandy Berro, "Obama Aide, Daley Pal Cash In on Chicago FBI Building Project," *Bettergov.org*, September 15, 2014, http:// www.bettergov.org/news/obama-aide-daley-pal-cash-in-on-chicago -fbi-building-project.

36 Penny Pritzker, U.S. Sec'y of Commerce, Office of Public Affairs, The "Open for Business" Agenda (November 13, 2013), https://www.commerce .gov/news/secretary-speeches/2013/11/open-business-agenda.

37 Daniel J. Sernovitz, "Alexandria's Carlyle Center Sells for $78 Million," *Washington Business Journal*, February 15, 2013, http://www.bizjournals .com/washington/breaking_ground/2013/02/alexandrias-carlyle-center -sells-for.html.

38 "Artemis Real Estate Partners' Emerging Manager Program and American Real Estate Partners Announce First Office Acquisition in Alexandria, Vir-

ginia," American Real Estate Partners, accessed July 14, 2017, http://www.americanrepartners.com/2013/03/artemis-real-estate-partners-emerging-manager-program-and-american-real-estate-partners-announce-first-office-acquisition-in-alexandria-virginia/.

39 "Lease Inventory," U.S. General Services Administration, accessed July 14, 2017, https://www.gsa.gov/portal/content/101840.

40 Allison Quinn-Redding, "Griffith Properties and Artemis Real Estate Partners Secure $30.39 Million Refi for Cambridge Office Bldg.," *CoStar*, April 5, 2016, http://www.costar.com/News/Article/Griffith-Properties-and-Artemis-Real-Estate-Partners-Secure-$3039-Million-Refi-for-Cabridge-Office-Bldg/181218.

41 "Lease Inventory."

42 Jon Peterson, "Ellis Partners Pays $34.9MM to Buy Office Building in San Rafael," *Registry*, October 30, 2014, http://news.theregistrysf.com/ellis-partners-pays-34–9mm-buy-office-building-san-rafael/.

43 "Lease Inventory."

44 Steve Brown, "Investors Nab Office Project in Shadow of Cowboys' Star Development in Frisco," *Dallas News*, August 31, 2016, https://www.dallasnews.com/business/business/2016/08/31/investors-nab-office-project-shadow-cowboys-star-development-frisco.

45 "Lease Inventory."

46 Onyx Equities, "Joint Venture of Onyx Equities and Artemis Real Estate Partners Acquires Mount Kemble Corporate Center in Morristown, N.J.," press release, October 9, 2013, http://www.thenewsfunnel.com/press-release/joint-venture-onyx-equities-and-artemis-real-estate-partners-acquires-mount-kemble-0.

47 "Avaya Relocates to Mount Kemble Corporate Center," *New Jersey Business*, April 21, 2016, https://njbmagazine.com/njb-news-now/avaya-relocates-mount-kemble-corporate-center/.

48 Advanced Data Search: Avaya, Contracts, Grants, Loans, Other Financial Assistance, for the Fiscal Years 2008–2017, *USASpending.gov* webpage, accessed July 14, 2017, https://www.usaspending.gov/Pages/AdvancedSearch.aspx?sub=y&ST=C,G,L,O&FY=2017,2016,2015,2014,2013,2012,2011,2010,2009,2008&A=0&SS=USA&k=avaya&pidx=8&SB=RN&SD=ASC.

49 CT Realty Investors, "CT Realty Investors, Artemis Real Estate Partners Acquire Huntington Beach Industrial Property," press release, August 5, 2013, http://www.marketwired.com/press-release/ct-realty-investors-artemis-real-estate-partners-acquire-huntington-beach-industrial-1817699.htm.

50 *Terms and Conditions of Sale*, Zodiac Services Americas LLC, December 17, 2015, http://www.zodiacaerospace.com/sites/default/files/content-images/docs-page-filiales/gtcs_us_vk_17.12.2015.pdf.

51 Advanced Data Search: Zodiac of North America, Contracts, for the Fiscal Years 2011–2017, to the Department of Commerce, *USASpending.gov* webpage, accessed July 14, 2017, https://www.usaspending.gov/Pages/AdvancedSearch.aspx?sub=y&ST=C&FY=2017,2016,2015,2014,2013,2012,2011&A=0&SS=USA&AA=1300&k=zodiac+of+north+america&SB=AD&SD=ASC.

52 White House Office of the Press Secretary, "FACT SHEET: White House Announces Doubling of TechHire Communities, and New Steps to Give More Students and Workers Tech Skills to Fuel the Next Generation of American Innovation," press release, March 9, 2016, https://obamawhitehouse.archives.gov/the-press-office/2016/03/09/fact-sheet-white-house-announces-doubling-techhire-communities-and-new.

53 Daniel J. Sernovitz, "Why Republic Properties Corp. is Willing to Take a Chance on Spec Development Near Capitol Hill," *Washington Business Journal*, June 25, 2014, http://www.bizjournals.com/washington/breaking_ground/2014/06/republic-properties-to-go-spec-in-noma.html.

54 Daniel J. Sernovitz, "This Spec Building on Capitol Hill is About to Land Two Tenants," *Washington Business Journal*, June 17, 2015, http://www.bizjournals.com/washington/breaking_ground/2015/06/this-spec-building-on-capitol-hill-is-about-to.html.

55 U.S. Department of Commerce, "Office of Legislative and Intergovernmental Affairs (OLIA)," accessed July 14, 2017, https://www.commerce.gov/doc/os/office-legislative-and-intergovernmental-affairs.

56 National Association of Counties, *Adopted Interim Policy Resolutions*, 2015 NACo Legislative Conference, accessed July 14, 2017, http://www.naco.org/sites/default/files/documents/NACo2015InterimPolicyResolutions.pdf.

57 U.S. Economic Development Administration, "U.S. Department of Commerce and the National League of Cities to Host Community Forums in Eight Cities to Help Support Underserved Communities," *Newsroom* (blog), March 17, 2016, https://www.eda.gov/archives/2016/news/blogs/2016/03/17/eda-nlc.htm.

58 Jonathan O'Connell, "Former GSA Chief Lands at $2 Billion Real Estate Firm," *Washington Post*, February 26, 2015, https://www.washingtonpost.com/news/digger/wp/2015/02/26/former-gsa-chief-lands-at-2-billion-real-estate-firm/?utm_term=.3eedfbacd766.

59 "Administrator, General Services Administration," on Political Appointee Project's website, accessed July 14, 2017, http://www.politicalappointeeproject.org/administrator-general-services-administration.html.

60 U.S. Economic Development Administration, "U.S. Commerce Secretary Announces $10 Million in Grants to Advance Innovation Across America," press release, March 30, 2015, https://www.eda.gov/archives/2016/news/press-releases/2015/03/30/ris.htm.

61 Amina Elahi, "Clean Energy Trust Awarded $250,000 Federal Grant for Prize Fund," *Chicago Tribune*, March 30, 2015, http://www.chicagotribune.com/bluesky/originals/chi-clean-energy-trust-penny-pritzker-amy-francetic-bsi-story.html.

62 Ibid.

CHAPTER 12: THE TRUMP PRINCELINGS

1 Drew Harwell and Anu Narayanswamy, "Trump's Global Business Empire Raises Concerns About Foreign Influence," *Chicago Tribune*, November 20, 2016, http://www.chicagotribune.com/news/nationworld/politics/ct-trump-global-business-20161120-story.html.

2 John Hayward, "Peter Schweizer: Trump Family Will Be Offered 'Sweetheart Deals' by Foreign Governments; Needs Transparency," *Breitbart*, December 9, 2016, http://www.breitbart.com/radio/2016/12/09/schweizer-trump-family-offered-sweetheart-deals-foreign-governments-needs-transparency/.

3 Norman Eisen and Peter Schweizer, "The Swamp is Deep, and Here Are Five Bipartisan Ways to Drain It," *Washington Post*, December 13, 2016, https://www.washingtonpost.com/opinions/the-swamp-is-deep-and-here-are-five-bipartisan-ways-to-drain-it/2016/12/13/720a747a-c096–11e6–897f-918837dae0ae_story.html.

4 Daniel Golden, "How Did 'Less Than Stellar' High School Student Jared Kushner Get into Harvard?" *Guardian*, November 18, 2016, https://www.theguardian.com/commentisfree/2016/nov/18/jared-kushner-harvard-donald-trump-son-in-law; Byron York, "The Sordid Case Behind Jared Kushner's Grudge Against Chris Christie," *Washington Examiner*, April 16, 2017, http://www.washingtonexaminer.com/byron-york-the-sordid-case-behind-jared-kushners-grudge-against-chris-christie/article/2620427; Jill Colvin and Vivian Salama, "Christie to Lead Trump Administration's Drug Addiction Task Force," *Chicago Tribune*, March 29, 2017, http://www.chicagotribune.com/news/nationworld/ct-christie-trump-drug-task-force-20170329-story.html.

5 Written Agreement Between Federal Deposit Insurance Corporation and Federal Reserve System, Docket No. 05–010-B-HC (Feb. 10, 2005), https://www.federalreserve.gov/boarddocs/press/enforcement/2005/20050210/attachment.pdf; York, "The Sordid Case Behind Jared Kushner's Grudge Against Chris Christie."

6 Adam Piore, "Behind the Record Deal for 666 Fifth Avenue," *Real Deal*, October 22, 2007, https://therealdeal.com/issues_articles/behind-the-record-deal-for-666-fifth-avenue/.

7 Devin Leonard, "Jared Kushner's Trump Card," *Bloomberg Businessweek*, May 5, 2016, https://www.bloomberg.com/features/2016-jared-kushner-trump-card.

8 "AIG/Morgan Venture Paying $1.9Bln for Kushner's Apartments," *Commercial Real Estate Direct*, June 26, 2007, http://www.crenews.com/general_news/northeast/aigmorgan-venture-paying-1.9bln-for-kushners-apartments.html; Charles V. Bagli, "At Kushner's Flagship Building, Mounting Debt and a Foundered Deal," *New York Times*, April 3, 2017, https://www.nytimes.com/2017/04/03/nyregion/kushner-companies-666-fifth-avenue.html?_r=0.

9 Bagli, "At Kushner's Flagship Building, Mounting Debt and a Foundered Deal"; Eliot Brown, "The Cost of Paying Top Dollar," *Wall Street Journal*, March 30, 2011, https://www.wsj.com/articles/SB10001424052748703461504576231132984824372.

10 Shannon Pettypiece, "Trump's Biggest Goals Put at Risk After Kushner Sucked into Probe," *Bloomberg*, June 1, 2017, https://www.bloomberg.com/news/articles/2017–06–01/trump-s-biggest-goals-at-risk-as-kushner-is-sucked-into-probe; Mark Landler and Maggie Haberman, "Jared Kushner is About to Plunge into Middle East Diplomacy," *New York Times*, June 19, 2017, https://www.nytimes.com/2017/06/19/us/politics/jared-kushner-mideast-diplomacy.html.

11 Mark Landler, "China Learns How to Get Trump's Ear: Through Jared Kushner," *New York Times*, April 2, 2017, https://www.nytimes.com/2017/04/02/us/politics/trump-china-jared-kushner.html.

12 Shi Jiangtao, "Has China Overplayed the Kushner Card in Dealing with the White House?" *South China Morning Post*, September 20, 2017, http://www.scmp.com/news/china/diplomacy-defence/article/2111921/has-china-overplayed-kushner-card-dealing-white-house.

13 Jill Abramson, "The Princeling in the West Wing," *New York Times*, May 10, 2017, https://www.nytimes.com/2017/05/10/opinion/jill-abramson-the-princeling-in-the-west-wing.html.

14 Associated Press, "Ivanka Trump's business ties in China are shrouded in secrecy," *Los Angeles Times*, September 26, 2017, http://www.latimes.com/business/la-fi-ivanka-trump-china-20170926-story.html; "Zhang Huarong," Center for China & Globalization website, accessed October 11, 2017, http://en.ccg.org.cn/people/zhang-huarong/; Adam Minter, "A Mystery at Ivanka Trump's Shoe Factory," *Bloomberg.com*, May 31, 2017, https://www.bloomberg.com/view/articles/2017–05–31/a-mystery-at-ivanka-s-shoe-factory.

15 Arash Massoudi and James Fontanella-Khan, "China's Anbang Agrees $6.5bn Hotel Deal with Blackstone," *Financial Times*, March 13, 2016, retrieved March 16, 2016, https://www.ft.com/content/fd6b524a-e92e-11e5–888e-2eadd5fbc4a4?mhq5j=e1.

16 Rebecca Lin, "Anbang Chief Wu Xiaohui Detained by Authorities," *Sino-US*, June 14, 2017, http://sino-us.com/10/13090594767.html.

17 Susanne Craig, Jo Becker, and Jesse Drucker, "Jared Kushner, a Trump In-Law and Adviser, Chases a Chinese Deal," *New York Times*, January 7, 2017,

https://www.nytimes.com/2017/01/07/us/politics/jared-kushner-trump-business.html; Michael Forsythe and Charles V. Bagli, "No Deal Between Kushners and Chinese Company Over Fifth Avenue Skyscraper," *New York Times*, March 19, 2017, https://www.nytimes.com/2017/03/29/nyregion/talks-end-between-kushners-and-chinese-company.html.

18 Keith Bradsher and Sui-Lee Wee, "Why Did China Detain Anbang's Chairman? He Tested a Lot of Limits," *New York Times*, June 14, 2017, https://www.nytimes.com/2017/06/14/business/china-anbang-wu-xiaohui-detained.html.

19 Ben Walsh, Ryan Grim, and Clayton Swisher, "Jared Kushner Tried and Failed to Get a Half-Billion-Dollar Bailout from Qatar," *Intercept*, July 10, 2017, https://theintercept.com/2017/07/10/jared-kushner-tried-and-failed-to-get-a-half-billion-dollar-bailout-from-qatar/.

20 Ibid.

21 Mark Perry, "Tillerson and Mattis Cleaning Up Kushner's Middle East Mess," *American Conservative*, June 27, 2017, http://www.theamericanconservative.com/articles/tillerson-and-mattis-cleaning-up-kushners-middle-east-mess/.

22 David Filipov et al, "White House, Russian Bank Differ on Jared Kushner Secret Meeting," *Chicago Tribune*, June 1, 2017, http://www.chicagotribune.com/news/nationworld/politics/ct-jared-kushner-russian-bank-meeting-20170601-story.html.

23 Katie Benner, "The other Kushner Brother's Big Bet," *New York Times*, January 13, 2017, https://www.nytimes.com/2017/01/13/technology/jared-kushner-brother-joshua-kushner-spotlight.html?_r=0.

24 Jared Kushner's Executive Branch Personnel Public Financial Disclosure Report, Form 278e (filed with the U.S. Office of Government Ethics on March 31, 2017), https://assets.documentcloud.org/documents/3534020/Jared-Kushner-Financial-Disclosure.pdf.

25 Sara Buhr, "OpenGov Picks Up $25 Million More and Adds Marc Andreessen to the Board," *TechCrunch*, October 15, 2015, https://techcrunch.com/2015/10/15/opengov-picks-up-25-million-more-and-adds-marc-andreessen-to-the-board/.

26 Steven Bertoni, "Oscar Health Gets $400 Million and a $2.7 Billion Valuation from Fidelity," *Forbes*, February 22, 2016, https://www.forbes.com/sites/stevenbertoni/2016/02/22/oscar-health-gets-400-million-and-a-2–7-billion-valuation-from-fidelity/#4b5367a53211; "Oscar," StartupFlux profile, accessed August 24, 2017, https://startupflux.com/companies/hioscar.com.

27 Jonathan Ferziger, "Israeli Company That Fenced In Gaza Angles to Help Build Trump's Mexico Wall," *Bloomberg*, January 29, 2017, https://www.bloomberg.com/news/articles/2017–01–29/israel-s-magal-pushes-for-mexico-wall-deal-as-trump-buoys-shares.

28 Leigh Kamping-Carder, "Kirsh Scions Buy Fourth Condo at Trump International," *Real Deal*, March 1, 2012, https://therealdeal.com/2012/03/01/kirsh-scions-buy-fourth-condo-at-trump-international/.

29 "Israeli Billionaire is Kushner's Partner on 15 Manhattan Buildings," *Real Deal*, April 26, 2017, https://therealdeal.com/2017/04/26/israeli-billionaire-is-kushners-partner-on-15-manhattan-buildings/.

30 Rebecca Ballhaus and Richard Rubin, "Jared Kushner Discloses Dozens More Assets in Revised Financial Filing," *Wall Street Journal*, July 21, 2017, https://www.wsj.com/articles/jared-kushner-discloses-dozens-more-assets-in-revised-financial-filing-1500678017; Jared Kushner's Executive Branch Personnel Public Financial Disclosure Report, OGE Form 278e.

31 Rosalind S. Helderman and Tom Hamburger, "Trump's Presidency, Overseas Business Deals and Relations with Foreign Governments Could All Become Intertwined," *Washington Post*, November 25, 2016, https://www.washingtonpost.com/politics/trumps-presidency-overseas-business-deals-and-relations-with-foreign-governments-could-all-become-intertwined/2016/11/25/d2bc83f8-b0e2–11e6–8616–52b15787add0_story.html?utm_term=.c60795439661.

32 Peter Schroeder, "Report: Trump Pressed Argentina's President About Stalled Building Project," *The Hill*, November 21, 2016, http://thehill.com/policy/finance/307050-report-trump-pressed-foreign-building-project-in-congratulatory-phone-call.

33 Andy Kroll and Russ Choma, "Businesswoman Who Bought Trump Penthouse Is Connected to Chinese Intelligence Front Group," *Mother Jones*, March 15, 2017; Jennifer Gould Keil, "$16M Trump penthouse sells to exec with ties to China," *New York Post*, February 28, 2017, http://nypost.com/2017/02/28/16m-trump-penthouse-sells-to-exec-with-ties-to-china/.

34 Jim Zarroli, "When Is a Deal Just a Deal? When Trump Sells a Property, It's Not Always Clear," *NPR.org*, March 22, 2017, https://www.npr.org/2017/03/22/521096579/when-is-a-deal-just-a-deal-when-trump-sells-a-property-its-not-always-clear.

35 "About Global Alliance Associates, Angela Chen-Managing Director," Global Alliance Associates website, gaa.lucita.org, accessed October 4, 2017, http://gaa.lucita.org/about_who_angela.shtml.

36 Ibid.

37 Kroll and Choma, "Businesswoman Who Bought Trump Penthouse Is Connected to Chinese Intelligence Front Group."

38 Ibid.; Bates Gill and James Mulvenon, "Chinese Military-Related Think Tanks and Research Institutions," *China Quarterly* 171 (September 2002): 617–624.

39 Associated Press, "Why Donald Trump's Washington, D.C. Hotel Is So Controversial," *Fortune*, December 7, 2016, http://fortune.com/2016/12/07/donald-trump-hotel-washington-dc/.

40 Jackie Northam, "Trump Said He Would Donate Foreign-Government Profits; Docs Suggest Limited Effort," *NPR.org*, May 25, 2017, http://www.npr.org/2017/05/25/529896841/trump-said-he-would-donate-foreign-government-profits-docs-suggest-limited-effor.

41 Greg Farrell and Caleb Melby, "Deutsche Bank in Bind Over How Much to Modify $300 Million Trump Debt," *Bloomberg*, March 27, 2017, https://www.bloomberg.com/news/articles/2017–03–27/deutsche-bank-in-bind-over-how-to-modify-300-million-trump-debt.

42 Susanne Craig, "Trump's Empire: A Maze of Debts and Opaque Ties," *New York Times*, August 20, 2016, https://www.nytimes.com/2016/08/21/us/politics/donald-trump-debt.html?_r=0.

43 "Strengths," BHR Partners, accessed August 29, 2017, (the screenshot of the site was captured October 4, 2014–February 11, 2017), https://web.archive.org/web/20141225022502/; http://en.bhrpe.com:80/_d276561105.htm. Note: the first source is the current (as of 7/19/17) Bohai website to show its relationship with Bank of China. The second is an archived site to show Bohai's relationship with Rosemont Seneca: http://www.boc.cn/en/investor/ir6/201504/t20150402_4830133.html.

44 Jeremy Venook, "Trump's Interests vs. America's, Dubai Edition," https://www.theatlantic.com/business/archive/2017/08/donald-trump-conflicts-of-interests/508382/.

45 Ibid.

46 Caroline Houck, "Tracking Trump's National-Security Conflicts of Interest," *Defense One*, December 17, 2016, http://www.defenseone.com/politics/2016/12/next-national-security-conflict-trumps-foreign-business-interests/133899/.

47 Louise Story and Stephanie Saul, "Stream of Foreign Wealth Flows to Elite New York Real Estate," *New York Times*, February 7, 2015, https://www.nytimes.com/2015/02/08/nyregion/stream-of-foreign-wealth-flows-to-time-warner-condos.html. Note: this is the first in the series of six articles on the subject.

48 "Trump Soho," *Real Deal*, accessed August 27, 2017, https://therealdeal.com/new-research/topics/property/trump-soho-hotel-condominium/.

49 Kevin G. Hall, "A Canadian Bankruptcy Adds to Trump's Legal Woes," *McClatchy DC*, January 10, 2017, retrieved May 30, 2017; Brian Kates and Rich Schapiro, "Donald Trump Pal Tevfik Arif Busted in Turkey for Allegedly Running Hooker Ring Aboard Yacht," *Daily News*, October 1, 2010, http://www.nydailynews.com/news/world/trump-pal-busted-allegedly-running-hooker-ring-yacht-article-1.191136.

50 Andrew Cave, "How to Get Back a Lost $10B: One Bank's Tale in Europe's Biggest Alleged Fraud," *Forbes*, February 6, 2017, https://www.forbes.com/sites/andrewcave/2017/02/06/how-to-get-back-a-lost-10bn-one-banks-tale-in-europes-biggest-alleged-fraud/#4c95e87316cd.

51 Tom Burgis, "Dirty Money: Trump and the Kazakh Connection," *Financial Times*, October 19, 2016, https://www.ft.com/content/33285dfa-9231-11e6-8df8-d3778b55a923?mhq5j=e1.

52 "Trump Picked Mafia-linked Stock Fraud Felon As Senior Adviser," *Chicago Tribune*, December 4, 2015, http://www.chicagotribune.com/news/nationworld/politics/ct-trump-felix-sater-felon-adviser-20151204-story.html; Charles V. Bagli, "Real Estate Executive With Hand in Trump Projects Rose from Tangled Past," *New York Times*, December 17, 2007, http://www.nytimes.com/2007/12/17/nyregion/17trump.html; Richard Behar, "Donald Trump and The Felon: Inside His Business Dealings with a Mob-Connected Hustler," *Forbes*, October 3, 2016, https://www.forbes.com/sites/richardbehar/2016/10/03/donald-trump-and-the-felon-inside-his-business-dealings-with-a-mob-connected-hustler/#75cf17002282; Rosalind Helderman and Tom Hamburger, "Former Mafia-linked figure describes association with Trump," *Washington Post*, May 17, 2016, https://www.washingtonpost.com/politics/former-mafia-linked-figure-describes-association-with-trump/2016/05/17/cec6c2c6-16d3-11e6-aa55-670cabef46e0_story.html?utm_term=.0261a4c81b5b.

53 Megan Twohey and Scott Shane, "A Back-Channel Plan for Ukraine and Russia, Courtesy of Trump Associates," *New York Times*, February 19, 2017, https://www.nytimes.com/2017/02/19/us/politics/donald-trump-ukraine-russia.html?action=click&contentCollection=Politics&module=RelatedCoverage®ion=Marginalia&pgtype=article; Victim's Rights Amendment: Testimony in support of H.J. Res. 45, Before the House Judiciary Committee, Subcomm. On the Constitution, 114th Cong. 70 (2015) (statement of Paul G. Cassell, Ronald N. Boyce Presidential Professor of Criminal Law), https://judiciary.house.gov/wp-content/uploads/2016/02/05012015-Cassell-Testimony.pdf, 70.

54 Matt Apuzzo and Maggie Haberman, "Trump Associate Boasted That Moscow Business Deal 'Will Get Donald Elected,'" *New York Times*, August 28, 2017, https://www.nytimes.com/2017/08/28/us/politics/trump-tower-putin-felix-sater.html; Carol D. Loennig, Tom Hamburger, and Rosalind S. Helderman, "Trump's Business Sought Deal on a Trump Tower in Moscow While He Ran for President," *Washington Post*, August 27, 2017, https://www.washingtonpost.com/politics/trumps-business-sought-deal-on-a-trump-tower-in-moscow-while-he-ran-for-president/2017/08/27/d6e95114-8b65-11e7-91d5-ab4e4bb76a3a_story.html?utm_term=.ec3903aa018a, Nexis.

55 Apuzzo and Haberman, "Trump Associate Boasted Moscow Business Deal 'Will Get Donald Elected.'"

56 Andrew Rice, "The Original Russia Connection," *New York Magazine*, August 3, 2017, http://nymag.com/daily/intelligencer/2017/08/felix-sater-donald-trump-russia-investigation.html; Leonnig, Hamburger, and

Helderman, "Trump's Business Sought Deal on a Trump Tower in Moscow While He Ran for President."

57 Rice, "The Original Russia Connection."

58 Charles V. Bagli, "Brass Knuckles Over 2 Broadway; M.T.A. and Landlord Are Fighting It Out Over Rent and Renovations," *New York Times*, August 9, 2000, http://www.nytimes.com/2000/08/09/nyregion/brass-knuckles-over-2-broadway-mta-landlord-are-fighting-it-over-rent.html.

59 Ben Schreckinger, "Trump's Mob-linked Ex-associate Gives $5,400 to Campaign," *Politico*, August 26, 2016, http://www.politico.com/story/2016/08/donald-trump-russia-felix-sater-227434; Ben Schreckinger, "The Happy-Go-Lucky Jewish Group That Connects Trump and Putin," *Politico*, April 9, 2017, http://www.politico.com/magazine/story/2017/04/the-happy-go-lucky-jewish-group-that-connects-trump-and-putin-215007.

60 Schreckinger, "The Happy-Go-Lucky Jewish Group That Connects Trump and Putin."

61 Zev Chafets, "The Missionary Mogul," *New York Times*, September 16, 2007, http://www.nytimes.com/2007/09/16/magazine/16Leviev-t.html; Michael Kranish, "Kushner Firm's $285 Million Deutsche Bank Loan Came Just Before Election Day," *Washington Post*, June 25, 2017, https://www.washingtonpost.com/national/kushner-firms-285-million-deutsche-bank-loan-came-just-before-election-day/2017/06/25/984f3acc-4f88–11e7-b064–828ba60fbb98_story.html?utm_term=.112889d7557b.

62 Caleb Melby and Keri Geiger, "Behind Trump's Russia Romance, There's a Tower Full of Oligarchs," *Bloomberg*, March 16, 2017, https://www.bloomberg.com/news/articles/2017–03–16/behind-trump-s-russia-romance-there-s-a-tower-full-of-oligarchs.

63 "About Sam Kislin," on Sam Kislin's website, accessed August 27, 2017, http://samkislin.weebly.com/.

64 Robert I. Friedman, *Red Mafiya: How the Russian Mob Has Invaded America* (Boston: Little, Brown, 2000), page number unknown, https://books.google.com/books?id=F1IqGCc7P8cC&printsec=frontcover&dq=red+mafyia&hl=en&sa=X&ved=0ahUKEwjZi5fIzpjVAhXNZiYKHYHCAfMQ6AEIKjAA#v=onepage&q=Sam%20Kislin&f=false; Harrah's Atlantic City, Order No. 00143 (New Jersey Office of the Attorney General, Department of Law and Public Safety division of Gaming Enforcement), http://www.nj.gov/oag/ge/docs/Rulings/2014/nov1_15/b10_caesarsresubmission.pdf.

65 Knut Royce, "FBI Tracked Alleged Russian Mob Ties of Giuliani Campaign Supporter," The Center for Public Integrity, August 20, 2000, https://www.ire.org/resource-center/stories/16211/.

66 Zephyr Teachout, "Trump Could Be the Most Corruptible President Ever," *Politico*, December 21, 2016, http://www.politico.com/magazine/story/2016/12/donald-trump-corruptible-214541.

67 Ibid.

68 Eisen and Schweizer, "The Swamp is Deep, and Here Are Five Bipartisan Ways to Drain It."

69 Conor Lynch, "Trump's Conflicts Are Unprecedented, but Not Unique: A Short History of Republican Corruption," *Salon*, February 8, 2017, http://www.salon.com/2017/02/08/trumps-conflicts-are-unprecedented-but-not-unique-a-short-history-of-republican-corruption-/.

CONCLUSION

1 U.S. Securities and Exchange Commission, "JP Morgan Chase Paying $264 Million to Settle FCPA Charges," press release, January 24, 2016, https://www.sec.gov/news/pressrelease/2016–241.html.

2 Luke Johnson and Ryan J. Reilly, January 21, 2014, "Bob McDonnell, Wife Indicted On Federal Corruption Charges For Accepting Gifts," Huffington Post, https://www.huffingtonpost.com/2014/01/21/bob-mcdonnell-corruption_n_4639480.html; Robert Barnes, "Supreme Court Overturns Corruption Conviction of Former Va. Governor McDonnell, Washington Post, June 27, 2016, https://www.washingtonpost.com/politics/supreme-court-rules-unanimously-in-favor-of-former-va-robert-f-mcdonnell-in-corruption-case/2016/06/27/38526a94-3c75-11e6-a66f-aa6c1883b6b1_story.html?utm_term=.a0f0105570f5

3 "George Bush's Letter Cautions Family on Conflicts of Interest," *New York Times*, April 17, 2015, https://www.nytimes.com/interactive/2015/04/17/us/politics/document-bush-letter-to-bush.html.

4 Ibid.

INDEX

INDEX

INDEX

ABOUT THE AUTHOR

PETER SCHWEIZER is the author of, among other books, *Clinton Cash*, *Extortion*, *Throw Them All Out*, and *Architects of Ruin*. His work on corruption has been featured throughout the media, including on *60 Minutes* and in the *New York Times*. He is the cofounder and president of the Government Accountability Institute, a team of investigative researchers and journalists committed to exposing crony capitalism, misuse of taxpayer monies, and other governmental corruption or malfeasance. He lives in Tallahassee, Florida.